BG Chemie

Toxicological Evaluations

1 Potential Health Hazards
of Existing Chemicals

Springer-Verlag
Berlin Heidelberg New York
London Paris Tokyo
Hong Kong Barcelona

Berufsgenossenschaft
der chemischen Industrie
Gaisbergstraße 11
D-6900 Heidelberg 1

ISBN 3-540-52577-7 Springer Verlag Berlin Heidelberg New York
ISBN 0-387-52577-7 Springer Verlag New York Berlin Heidelberg

Library of Congress Cataloging-in-Publication Data. Toxicological evaluations/BG Chemie p. cm. Includes bibliographical references. Includes index. Contents: 1. Potential health hazards of existing chemicals.
 ISBN 3-540-52577-7 (v. 1: alk. paper)
 ISBN 0-387-52577-7 (v. 1: alk. paper)
1. Toxicity testing. I. Berufsgenossenschaft der Chemischen Industrie. [DNLM: 1. Hazardous Substances – toxicity.   WA 465 T7549]   RA 1199. T678   1990   815.9'07–dc20   DNLM/DLC

This work is subject to copyright. All rights are reserved, whether the whole or part of the material is concerned, specifically the rights of translation, reprinting, reuse of illustrations, recitation, broadcasting, reproduction on microfilms or in other ways, and storage in data banks. Duplication of this publication or parts thereof is only permitted under the provisions of the German Copyright Law of September 9, 1965, in its current version, and a copyright fee must always be paid. Violations fall under the prosecution act of the German Copyright Law.

© Springer-Verlag Berlin Heidelberg 1990
Printed in Germany

The use of registered names, trademarks, etc. in this publication does not imply, even in the absence of a specific statement, that such names are exempt from the relevant protective laws and regulations and therefore free for general use.

Liability: The content of this document has been prepared and reviewed by experts on behalf of BG Chemie with all possible care and from the available scientific information. It is provided for information only. BG Chemie cannot accept any responsibility of liability and does not provide a warranty for any use of interpretation of the material contained in the publication.

Typesetting, printing and bookbinding: Brühlsche Universitätsdruckerei, Giessen
2152/3130-543210 – Printed on acid-free paper

Preface

As part of its programme for the prevention of health hazards caused by industrial work substances, the Berufsgenossenschaft der Chemischen Industrie began in 1977 to investigate the toxicity of those existing substances which are in widespread use, have many different applications and are suspected of being possibly dangerous to health, in particular of having long-term effects on health. It is hoped by means of this testing to close gaps in our knowledge and to increase the scientific validity of the required risk assessments. The results of the toxicological investigations carried out by the Berufsgenossenschaft der Chemischen Industrie, and the resulting substance assessments have been published in West Germany since 1987 in the form of "Toxicological Evaluations".

In order to make this useful information internationally available, the "Toxicological Evaluations" are now being published in English. This first volume contains individual evaluations of 21 substances. The publication of further individual evaluations and, if necessary, reassessments of previously published evaluations is planned.

The Berufsgenossenschaft der Chemischen Industrie hopes that, for many people working in the chemical industry, this information will be of practical help in assessing hazards to health in the workplace.

Contents

Name of substance	CAS-No.	Page
Introduction		1
Annex 1		9
Annex 2		12
Annex 3		13
Annex 4		15
Dicyclopentadiene	77-73-6	23
2-Methylpropanol-1	78-83-1	43
Chloroacetamide	79-07-2	59
4-Chloro-2-nitroaniline	89-63-4	75
Ethylene thiourea	96-45-7	81
Gamma-Butyrolactone	96-48-0	133
5-Nitroanisidine	99-59-2	155
N,N'-Di-sec.-butyl-p-phenylenediamine	101-96-2	165
m-Cresidine	102-50-1	173
2-Ethylhexanol	104-76-7	181
Butynediol	110-65-6	207
Diethylene glycol	111-46-6	217
1,4-Dicyanobutane	111-69-3	251
Dimethyl therephthalate	120-61-6	265
3-Methylbutanol-1	123-51-3	283
Tributyl phosphate	126-73-8	297
Trimethylphenyl-ammonium chloride	138-24-9	311
Chloroformic acid ethyl ester	541-41-3	319
Manganese dioxide	1313-13-9	327
Chemical index (see also Annex 4)		339

Introduction

It is estimated that, at present, about 100 000 existing chemicals are marketed for various purposes. Between 5 and 10% of these substances are regarded by the OECD as possibly dangerous. Only a small proportion, however, of these 100 000 existing chemicals are of technical significance. According to a survey made by the Chemical Industry Association (VCI)[1] of the Federal Republic of Germany (FRG), only about 4600 substances are produced in quantities greater than 10 tons annually. Most of the remaining substances are laboratory chemicals or products manufactured or delivered in reponse to individual orders.

Whereas substances which, since 1981, have been placed on the market for the first time ("new chemicals") are required by the Act on Chemicals[2] to be submitted to evaluation under special conditions, this is necessary only in certain cases for existing chemicals in order to comply with specific regulations.

The Employment Accident Insurance Fund of the Chemical Industry – Berufsgenossenschaft der Chemischen Industrie (BG Chemie) – was already concerned with the testing of industrial substances for health-endangering properties long before the Act on Chemicals came into force. Its contribution in this field is made available in these **Toxicological Evaluations** and in further material to be published. The work of the BG Chemie in this field has now been incorporated into the Federal Government's scheme for systematic testing and evaluation of existing chemicals.

In 1977 the management of the BG Chemie decided to give particular priority to the prevention of damage to health through industrial products. For this purpose it set up the

Programme for the prevention of health hazards caused by industrial substances.

In addition to other activities, the BG Chemie has begun to investigate, particularly for long-term effects, chemicals whose possible hazards to health are so far only suspected, so that additional

---

[1] Verband der Chemischen Industrie (VCI)
[2] Chemikaliengesetz

Introduction

safety measures can promptly be introduced into the factories when they are indicated by the results of these investigations.

For testing purposes, these substances have been compiled in a list. The prerequisites for inclusion in this list are as follows:

1. The substances in question should be those used in industry in the manufacture of primary, intermediate, final or auxiliary products.
2. The volume of production and the method of handling the substances are also relevant considerations.
3. Special priority is given to substances that are destined for the consumer sector.
4. There should be evidence of a potential risk to health. This could include, for example, experience gathered in the workplace or by occupational physicians, unconfirmed indications in the literature or a similarity of the chemical structure to other substances that have proved to be hazardous (e.g. alkylating agents, aromatic amines).
5. Basically, the list compiled by the BG Chemie will not include substances for which threshold limit values have been fixed or which have proved to be carcinogenic in man or in animal studies.
6. In order to avoid unnecessary duplication of work, BG Chemie will not study substances which are known to be the subject of investigations already being carried out by other national or international groups, provided these studies pursue objectives similar to those of the BG Chemie.

Substances are proposed by the member companies and by the technical supervision section of the BG Chemie as well as by other persons or organizations, e.g. the Federal Agency for Occupational Safety[3]. The first substances put forward as a consequence of this procedure have already been submitted for evaluation.

In assessing individual substances, the BG Chemie is supported by a scientific advisory committee (see Annex 1), which includes experienced toxicologists, occupational physicians and chemists. Representatives of the Federal Institute for Occupational Safety and the Federal Health Office[4] also participate. The decision-making

---

[3] Bundesanstalt für Arbeitsschutz (BAU)
[4] Bundesgesundheitsamt

body for all these activities is the Principles and Substances Programme Committee (Annex 2).

After a substance has been included in the list, a painstaking investigation must be undertaken to decide whether that particular substance poses a health hazard to employees.

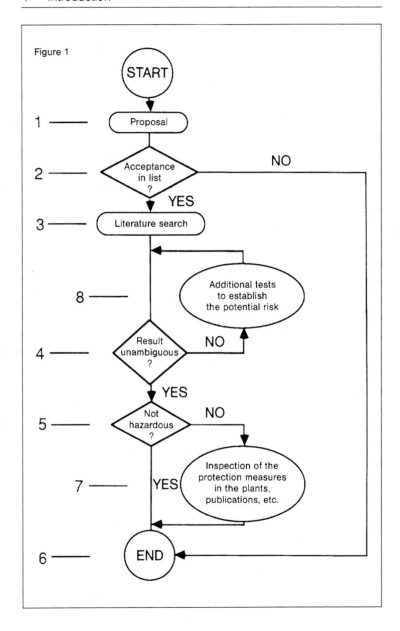

Figure 1

Introduction

The full procedure is as follows:

- Proposal for acceptance in the BG Chemie list of substances (1).
- Decision on whether or not to accept the substance (2).
- After acceptance in the list, a toxicological institute will conduct a search and review of the literature to see if there are any scientific publications or other study findings that report on acute or chronic toxicity, or on teratogenic, genotoxic or carcinogenic properties (3).
- On the basis of the literature search and review, a decision will be made as to whether a conclusive evaluation of the chemical in question is possible (4).
- If this shows that the substance is not hazardous (5), studies on this chemical can be concluded (6).
- If the substance studied has properties hazardous to health, the BG Chemie will check the safety precautions in the organizations where this chemical is used and, if necessary, will ensure that adequate improvements are made. At the same time, these results wil be published and made available to interested and affected parties; for example, to the Commission for the Investigation of Health Hazards of Chemical Compounds in the Work Area[5].
- If the literature search does not permit a conclusive evaluation of the substance (4), further studies must be conducted to determine the potential risk (8).
- Once these studies have been completed, a decision must be taken as to whether the available documentation is sufficient for a final evaluation (4).
- If the results are conclusive (4), no further tests are necessary. If on the other hand the chemical does pose risks, appropriate measures must be taken (7).
- However, if the results are inconclusive (4), additional tests will be necessary to assess the potential risk (8). The procedures to be followed will be decided on a case-to-case basis.

---

[5] Senatskommission zur Prüfung gesundheitsschädlicher Arbeitsstoffe (MAK Kommission)

6 Introduction

So far, the following tasks have been accomplished:

- Some 3800 substances were subjected to preliminary investigations to ascertain whether they fulfilled the above-mentioned prerequisites for inclusion in the list.
- The question of acceptance of 412 of the substances has been discussed intensively by the advisory committee, in some cases repeatedly.
- 279 substances have been placed on the List of substances (see Annex 4).
- In the meantime, literature searches have been conducted on 150 of these substances.
- 143 substances have been discussed by the scientific advisory committee.
- In the case of 88 substances, the discussions have been provisionally completed.
- For a number of substances the literature searches have produced no clear-cut results. Hence, a decision has been made to pursue the question further by conducting investigations to ascertain the potential risk, especially in terms of teratogenic, embryotoxic, neurotoxic, genotoxic or carcinogenic effects. For 83 substances, tests – in some cases several tests (244 altogether) – have been planned, commissioned or already completed. For three substances, long-term studies of carcinogenicity have also been planned.

On the basis of the work so far completed, evidence exists that the following 9 substances in the list

- N-Methyl-bis(2-chloroethyl-amine) [CAS-No. 51-75-2]
- Vinylidene-fluoride [CAS-No. 75-38-7]
- Diethylcarbamylchloride [CAS-No. 88-10-8]
- 5-Chloro-2-aminotoluene [CAS-No. 95-69-2]
- Benzotrichloride [CAS-No. 98-07-7]
- 4-Nitro-2-aminotoluene [CAS-No. 99-55-8]
- 1,2-Butyleneoxide [CAS-No. 106-88-7]
- 3,3'-Dimethyl-4,4'diaminodiphenyl methane [CAS-No. 838-88-0]
- 2-Chloroacrylonitrile [CAS-No. 920-37-6]

are carcinogenic.

In each individual case, the Technical Supervisory Officer has investigated the use of these substances on the plant premises and, wherever necessary, has introduced improvements to the occupational safety measures.

In order to make these results available to all interested parties, and in particular to specialists, the BG Chemie began in 1987 to publish the **Toxicological Evaluations** in German. In all, 57 such evaluations have now been published. They have been prepared on the basis of collaboration by scientists, toxicological research institutes (see Annex 3) and the secretariat of the advisory committee, which also undertook the work of sub-editing (see Annex 1).

These evaluations are based on documentation found in the scientific literature, on experimental studies commissioned by the BG Chemie and on experience gathered by the staff of the specialised institutes commissioned to make the toxicological evaluations, as well as by the members of the advisory committee. The members of this committee are in agreement with the toxicological evaluations submitted. The BG Chemie's Principles and Substances Programme committee has also accepted these evaluations.

The increasing international interest in the **Toxicological Evaluations** has led to the preparation of this publication in English, the first volume of which is presented now.

With its programme for the prevention of health hazards caused by industrial substances, the BG Chemie intends to ensure that the risks for employees in contact with dangerous industrial substances are recognized as early as possible and, if necessary, reduced by examining the situation at the workplace as well as by introducing improvements. The BG Chemie is grateful to all persons named in this report, without whose active cooperation the research could not have been efffectively pursued. We wish to thank also the many persons, who cannot be named individually here, who have contributed their time and energy to the success of these endeavours.

Annex 1

Members of the Advisory Committee

Dr. G. Fleischhauer (Chairman)
Berufsgenossenschaft der chemischen Industrie
(Employment Accident Insurance Fund of the Chemical Industry)
Postfach 101480, 6900 Heidelberg

Dr. L. Dix
Ciba-Geigy AG, Werksärztlicher Dienst (Works Medical Service),
Betriebsstätte Grenzach (Plant site Grenzach)
Postfach 1120, 7889 Grenzach-Wyhlen 1

Professor Dr. Dr. H.-P. Gelbke
BASF Aktiengesellschaft, Abteilung Toxikologie
(Department of Toxicology),
ZNT – Z 470
6700 Ludwigshafen

Professor Dr. D. Henschler
Institut für Pharmakologie und Toxikologie der Universität Würzburg
(Institute of Pharmacology and Toxicology
at the University of Würzburg)
Versbacher Strasse 9, 8700 Würzburg

Professor Dr. H. Herrmann
Jürgensallee 44 m, 2000 Hamburg 52

Dr. R. Jung
Hoechst AG – Abteilung Toxikologie (Department of Toxicology)
6230 Frankfurt 80

Dr. H. Jungen
Deutsche Wissenschaftliche Gesellschaft für Erdöl, Erdgas und Kohle e.V. (German Scientific Association of Oil, Gas and Coal)
Steinstrasse 7, 2000 Hamburg 1

Professor Dr. D. Kayser
Bundesgesundheitsamt (Federal Health Office)
Postfach 330013, 1000 Berlin 33

Dr. U. Korallus
Ärztliche Abteilung (Former Chief Works Physician), Bayer AG
5090 Leverkusen-Bayerwerk

Dr. E. Löser
Bayer AG – Institut für Toxikologie (Department of Toxicology)
Friedrich-Ebert-Strasse 217, 5600 Wuppertal 1

Professor Dr. A. Manz
Alter Achterkamp 61, 2070 Großhansdorf 2

Professor Dr. H. G. Miltenburger
Institut für Zoologie (Institute of Zoologie),
Technische Hochschule Darmstadt
Schnittspahnstrasse 3, 6100 Darmstadt

Direktor Dr. H. G. Peine
BASF Aktiengesellschaft
Bereich Umweltschutz und Arbeitssicherheit
(Department of Environmental Protection and Occupation Safety)
6700 Ludwigshafen

Professor Dr. R. Preussmann
Deutsches Krebsforschungszentrum
(West German Cancer Research Center)
Institut für Toxikologie und Chemotherapie
Im Neuenheimer Feld 280, 6900 Heidelberg

Dr. N. Rupprich
Bundesanstalt für Arbeitsschutz
(Federal Agency for Occupational Safety)
Postfach 170202, 4600 Dortmund 17

Direktor Dr. H.-K. Schäfer
Initiative Umweltrelevante Altstoffe
(Initiative on existing chemicals of environmental relevance)
Kennedyallee 93, 6000 Frankfurt 70

Dr. H. Schnierle
Hoechst AG – Sicherheitsüberwachung (Safety Monitoring)
Bau C 769
6230 Frankfurt 80

Dr. H. Schweinfurth
Schering AG, Industrie-Chemikalien (Industrial Chemicals)
Waldstrasse 14, 4709 Bergkamen

Permanent Guest

Dr. H.-J. Klimisch
BASF Aktiengesellschaft, Abteilung Toxikologie
(Department of Toxicology)
ZNT – Z 470
6700 Ludwigshafen

Secretariat of the Advisory Committee

Frau Dr. M. Beth Dr. H. Th. Hofmann
Frau G. Haass Dr. W. Huber
Dr. K.-G. Heimann Dr. J. Oberhansberg
Frau Ch. Heumann Dr. H. Zeller

Adress: Berufsgenossenschaft der chemischen Industrie
(Employment Accident Insurance Fund of the Chemical Industry)
Postfach 101480, 6900 Heidelberg 1

Annex 2

Members of the Principles and Substances Program Committee

Wolfgang Schultze
Industriegewerkschaft Chemie-Papier-Keramik
(Chemicals-Paper-Ceramics Trade Union)
Postfach 3047, 3000 Hannover

Eberhard Lechelt
Industriegewerkschaft Chemie-Papier-Keramik
(Chemicals-Paper-Ceramics Trade Union)
Postfach 3047, 3000 Hannover

Dr. Karl Molitor
Bundesarbeitgeberverband Chemie e.V.
(West German Employers' Association)
Postfach 1280, 6200 Wiesbaden

Professor Dr. Eberhard Weise
Bayer AG, 5090 Leverkusen-Bayerwerk

Chairman of the Board
of the Berufsgenossenschaft der chemischen Industrie
(Employment Accident Insurance Fund of the Chemical Industry)
Dr. Detlef Dibbern
BASF AG, 6700 Ludwigshafen

Vice-Chairman of the Board
of the Berufsgenossenschaft der chemischen Industrie
(Employment Accident Insurance Fund of the Chemical Industry)
Richard Siegler
Höchster Strasse 62, 6144 Groß-Umstadt

Managing director
of the Berufsgenossenschaft der chemischen Industrie
(Employment Accident Insurance Fund of the Chemical Industry)
Direktor Hanswerner Lauer
Berufsgenossenschaft der chemischen Industrie
Postfach 101480, 6900 Heidelberg 1

Annex 3

In addition to the members of the advisory committee and the secretariat the following persons were involved in getting by working for the Toxicological Evaluations:

Professor Dr. H. Marquardt
Direktor der Abteilung für allgemeine Toxikologie
Universitätskrankenhaus Eppendorf
Grindelallee 117, 2000 Hamburg 13

Dr. B. Görlitz
Fraunhofer-Institut für Toxikologie und Aerosolforschung
Institutsteil Hannover
Nikolai-Fuchs-Strasse 1, 3000 Hannover 61

Frau Dr. H. Habs
Fraunhofer-Institut für Toxikologie und Aerosolforschung
Institutsteil Hannover
Nikolai-Fuchs-Strasse 1, 3000 Hannover 61

Dr. M. Habs
Fraunhofer-Institut für Toxikologie und Aerosolforschung
Institutsteil Hannover
Nikolai-Fuchs-Strasse 1, 3000 Hannover 61

Professor Dr. F. Leuschner
Laboratorium für Pharmakologie und Toxikologie
Redderweg 8, 2104 Hamburg 92

Dr. J. Leuschner
Laboratorium für Pharmakologie und Toxikologie
Redderweg 8, 2104 Hamburg 92

Mrs. T. Russel
(Translation) British Industrial Biological Research Association
Woodmansterne Road, Surrey SM5 4DS
Great Britain

Annex 4

List of substances compiled by the Berufsgenossenschaft der chemischen Industrie

Name of substance	CAS-No.
N-Methyl-bis(2-chloroethyl)amine	51-75-2
Acetamide	60-35-5
N,N′-Diphenyl-p-phenylenediamine	74-31-7
Chloromethane	74-87-3
Vinyl fluoride	75-02-5
Bromoform	75-25-2
Vinylidene fluoride	75-38-7
Methyldichlorosilane	75-54-7
Trimethylmonochlorosilane	75-77-4
Dicyclopentadiene	77-73-6
Isoprene	78-79-5
2-Methylpropane-1-ol	78-83-1
Methacroleine	78-85-3
Methylvinylketone	78-94-4
Trichloroethene	79-01-6
Chloroacetyl chloride	79-04-9
Chloroacetamide	79-07-2
Acrylic acid	79-10-7
Chloroacetic acid	79-11-8
Chlorocarbonic acid methylester	79-22-1
Dichloroacetyl chloride	79-36-7
Methacrylamide	79-39-0
Methacrylic acid	79-41-4
Dichlorodiphenylsilane	80-10-4
Benzenesulfonic acid hydrazide	80-17-1
2-Nitro-1,3-dimethylbenzene	81-20-9
3-Nitro-1,2-dimethylbenzene	83-41-0
Anthraquinone	84-65-1
1-Nitronaphthalene	86-57-7
7-Amino-4-hydroxy-2-naphthalene sulfonic acid	87-02-5
Diethylcarbamyl chloride	88-10-8
o-Chlorobenzotrifluoride	88-16-4

Name of substance	CAS-No.
o-Nitrotoluene	88-72-2
o-Chloronitrobenzene	88-73-3
o-Nitroaniline	88-74-4
2-(1-Methylpropyl)-1-hydroxy-4,6-dinitrobenzene	88-85-7
Nitro-p-xylene	89-58-7
2-Nitro-4-methylaniline	89-62-3
4-Chloro-2-nitroaniline	89-63-4
4-Nitro-1,3-dimethylbenzene	89-87-2
N-Phenyl-1-naphthylamine	90-30-2
6-Amino-4-hydroxy-2-naphthalene sulfonic acid	90-51-7
o-Phthalodinitrile	91-15-6
Quinoline	91-22-5
4-Nitro-4′-aminodiphenylamine-2-sulfonic acid	91-29-2
Diphenyl	92-52-4
o-Chlorotoluene	95-49-8
o-Chloroaniline	95-51-2
2-Fluorotoluene	95-52-3
2,4-Xylidine	95-68-1
5-Chloro-2-aminotoluene	95-69-2
Chloro-p-xylene	95-72-7
2,4-Dichlorotoluene	95-73-8
3,4-Dichlorotoluene	95-75-0
3,4-Dichloroaniline	95-76-1
4-Chloro-2-aminotoluene	95-79-4
2,4-Toluylenediamine	95-80-7
2,5-Dichloro aniline	95-82-9
2,4,5-Trichlorophenol	95-95-4
2-Butanone oxime	96-29-7
Chloroacetic acid methylester	96-34-4
Ethylenethiourea	96-45-7
γ-Butyrolactone	96-48-0
1-Chloro-2,4-dinitrobenzene	97-00-7
N-(2,4-Dimethylphenyl)-3-oxobutanoic acid amide	97-36-9
1,3-Di-o-tolylguanidine	97-39-2
2-Methoxy-4-nitroaniline	97-52-9
o-Aminoazotoluene	97-56-3
Tetramethyl thiuram monosulfide	97-74-5
Benzotrichloride	98-07-7
Benzotrifluoride	98-08-8

Name of substance	CAS-No.
m-Chlorobenzotrifluoride	98-15-7
p-t-Butylphenol	98-54-4
p-t-Butylbenzoic acid	98-73-7
Benzalchloride	98-87-3
Benzoylchloride	98-88-4
Nitrobenzene	98-95-3
m-Nitrotoluene	99-08-1
m-Nitroaniline	99-09-2
4-Nitro-1,2-dimethylbenzene	99-51-4
5-Nitro-2-aminotoluene	99-52-5
4-Nitro-2-aminotoluene	99-55-8
5-Nitro-o-anisidine	99-59-2
N,N-Dimethyl-p-phenylenediamine	99-98-9
p-Nitrotoluene	99-99-0
4-Nitrophenol	100-02-7
Terephthalic acid	100-21-0
Benzylchloride	100-44-7
p-Aminodiphenylamine	101-54-2
Dioctyldiphenylamine	101-67-7
Dicyclohexylamine	101-83-7
N,N'-Di-sec-butyl-p-phenylenediamine	101-96-2
N-Phenyl-3-oxobutanoic acid amide	102-01-2
1,3-Diphenylguanidine	102-06-7
2-Amino-5-methoxytoluene	102-50-1
Triethanolamine	102-71-6
2-Ethylhexyl acrylate	103-11-7
Phenylisocyanate	103-71-9
Dimethylbenzylamine	103-83-3
4-Chlorophenylisocyanate	104-12-1
2-Ethylhexanol	104-76-7
p-Nitrosophenol	104-91-6
Chloroacetic acid ethyl ester	105-39-5
Carbonic acid diethyl ester	105-58-8
2,4-Dimethylphenol	105-67-9
Di-2-ethylhexylamine	106-20-7
p-Chlorotoluene	106-43-4
p-Chloroaniline	106-47-8
p-Toluidine	106-49-0
Diethylene glycol bis-chloroformic ester	106-75-2

Name of substance	CAS-No.
1,2-Butyleneoxide	106-88-7
Allyl chloride	107-05-1
Propargylalcohol	107-19-7
Ethyleneglycol	107-21-1
Glyoxal	107-22-2
Vinylmethylether	107-25-5
m-Chlorotoluene	108-41-8
3-Chloroaniline	108-42-9
3,5-Dimethylphenol	108-68-9
Cyanuric acid chloride	108-77-0
Cyanuric acid	108-80-5
Toluene	108-88-3
Isopropyl ethylene glycolether	109-59-1
Chlorocarbonic acid propylester	109-61-5
Tetrahydrofurane	109-99-9
Tetrahydrothiophene	110-01-0
1,4-Butanediol	110-63-4
2-Butyne-1,4-diol	110-65-6
Trioxane	110-88-3
Piperidine	110-89-4
Diisopropanolamine	110-97-4
Diethylenetriamine	111-40-0
Diethanolamine	111-42-2
Diethyleneglycol	111-46-6
1,4-Dicyanobutane	111-69-3
Diethylene glycol methylether	111-77-3
Diethylene glycol ethylether	111-90-0
Triethylenetetramine	112-24-3
2-Methylpropene	115-11-7
2-Methyl-3-butyn-2-ol	115-19-5
Tris(2-chloroethyl)-phosphate	115-96-8
Isatoic acid anhydride	118-48-9
4,4′-Methylene-bis(2,6-di-tert-butylphenol)	118-82-1
o-Chlorobenzoic acid	118-91-2
3-Amino-4-methoxybenzanilide	120-35-4
Terephthalic acid dimethylester	120-61-6
2,4-Dichlorophenol	120-83-2
Trimethyl phosphite	121-45-9
m-Chloronitrobenzene	121-73-3

Name of substance	CAS-No.
Triethyl phosphite	122-52-1
Phenylglycidylether	122-60-1
Hydrazobenzene	122-66-7
2-Ethylhexanal	123-05-7
Propionaldehyde	123-38-6
3-Methylbutane-1-ol	123-51-3
Diazene dicarboxamide	123-77-3
Dioxane	123-91-1
Isobornylacetate	125-12-2
Triisobutyl phosphate	126-71-6
Tributyl phosphate	126-73-8
Dimethylacetamide	127-19-5
1,4-Diaminoanthraquinone	128-95-0
1,4-Naphthoquinone	130-15-4
1-Naphthylamine	134-32-7
2-Naphthol	135-19-3
Thiram	137-26-8
Trimethylphenylammoniumchloride	138-24-9
3-Oxobutanoic acid ethyl ester	141-97-9
n-Heptane	142-82-5
2-Mercaptobenzothiazole	149-30-4
Triethylenediamine	280-57-9
Imidazole	288-32-4
4-Bromofluorobenzene	460-00-4
Fluorobenzene	462-06-6
Dicyclohexylcarbodiimide	538-75-0
Chlorocarbonic acid ethylester	541-41-3
β-Methallyl chloride	563-47-3
2,6-Dimethylphenol	576-26-1
Mercaptobenzimidazole	583-39-1
Chlorocarbonic acid butylester	592-34-7
α-Chloropropionic acid	598-78-7
1,2-Dichlorobutane	616-21-7
Maleic acid dimethylester	624-48-6
Propargylchloride	624-65-7
3-Chloropropanoic acid chloride	625-36-5
Methoxy acetic acid	625-45-6
Manganese (II) acetate	638-38-0
2,3,5-Trimethyl hydroquinone	700-13-0

Name of substance	CAS-No.
Diethyl phosphite	762-04-9
1,4-Dichlorobutene-2	764-41-0
3,3'-Dimethyl-4,4'-diaminodiphenylmethane	838-88-0
Dimethylhydrogen phosphite	868-85-9
2-Chloroacrylonitrile	920-37-6
N-Methylolmethacrylamide	923-02-4
Trimethylquinone	935-92-2
Diphenyl 2-ethylhexyl phosphate	1241-94-7
Manganesedioxide	1313-13-9
Sodium sulfide (anhydrous)	1313-82-2
Sodium sulfide (nonahydrate)	1313-84-4
Antimony (V) oxide	1314-60-9
Antimony (V) sulfide	1315-04-4
Copper-Phthalocyanine green	1328-53-6
Tricresyl phosphate (mixed isomers)	1330-78-5
Antimony (III) sulfide	1345-04-6
tert-Butylacrylate	1663-39-4
Dimethylaminopropionitrile	1738-25-6
2-Amino-6-methoxybenzothiazole	1747-60-0
1,2-Diaminoanthraquinone	1758-68-5
4,4'-Diaminodicyclohexylmethane	1761-71-3
p-Nitrocumene	1817-47-6
Dimethyloldihydroxyethylene urea	1854-26-8
2,4-Dinitro-N-methylaniline	2044-88-4
o-Chlorobenzotrichloride	2136-89-2
Diglycidylether	2238-07-5
1,5-Naphthylenediamine	2243-62-1
2,3,4-Trichlorobutene-1	2431-50-7
Aminoguanidinebicarbonate	2582-30-1
Ethylene glycol mono propyl ether	2807-30-9
Glycidyl trimethyl ammonium chloride	3033-77-0
Ethene sulfonic acid, sodium salt	3039-83-6
(3-Chloro-2-hydroxypropyl)trimethyl ammonium chloride	3327-22-8
1-(2,4-Dinitrophenylazo)-2-naphthol	3468-63-1
2,4-Dichloro-6-methoxy-1,3,5-triazine	3638-04-8
4-Chlorobutanoic acid chloride	4635-59-0
N,N-Dicyclohexyl-2-benzothiazole sulfenamide	4979-32-2
p-Chlorobenzotrichloride	5216-25-1
Hydroxylamine hydrochloride	5470-11-1

Annex 4   21

Name of substance	CAS-No.
Isobutylidenediurea	6104-30-9
Manganese (II) acetate (tetrahydrate)	6156-78-1
2,5-Dimethoxy-4-chloroaniline	6358-64-1
o-Nitrocumene	6526-72-3
Gallium	7440-55-3
Aluminum chloride	7446-70-0
Antimony (V) chloride	7647-18-9
Thioglycolic acid 2-ethylhexylester	7659-86-1
Sodium hypochlorite	7681-52-9
Aluminum bromide	7727-15-3
Ammoniumperoxidisulfate	7727-54-0
Manganese (II) chloride	7773-01-5
Hydrogen sulfide	7783-06-4
Antimony (III) fluoride	7783-56-4
Antimony (V) fluoride	7783-70-2
Aluminum chloride hexahydrate	7784-13-6
Aluminum nitrate nonahydrate	7784-27-2
Manganese (II) sulfate	7785-87-7
Ammoniumbichromate	7789-09-5
Antimony (III) bromide	7789-61-9
Chloro sulfonic acid	7790-94-5
Sulfuryl chloride	7791-25-5
Hydroxylamine	7803-49-8
Antimony (III) chloride	10025-91-9
Manganese (II) sulfate (monohydrate)	10034-96-5
Hydroxylamine sulfate	10039-54-0
Aluminum sulfate	10043-01-3
Hydroxylamine hydrogene sulfate	10046-00-1
Manganese (II) sulfate (tetrahydrate)	10101-68-5
Manganese (II) nitrate	10377-66-9
Sodium bichromate	10588-01-9
Dimethylaminosulfochloride	13360-57-1
Manganese (II) chlorid (tetrahydrate)	13446-34-9
Aluminum nitrate	13473-90-0
Sodium tetrafluoroborate	13755-29-8
Ammonium tetrafluoroborate	13826-83-0
Potassium tetrafluoroborate	14075-53-7
Zinc diethyl dithiocarbamate	14324-55-1
Zinc ethylphenyl dithiocarbamate	14634-93-6

Name of substance	CAS-No.
p-(2,4-Dichlorophenoxy)aniline	14861-17-7
Sodium hexafluoroaluminate	15096-52-3
Sodium bisulfide	16721-80-5
Tetrafluoro boric acid	16872-11-0
Di-t-butyl tin chloride	19429-30-2
Manganese (II) chloride (dihydrate)	20603-88-7
Manganese (II) nitrate (tetrahydrate)	20694-39-7
Ethylene glycol mono propyl ether acetate	20706-25-6
Tellurium diethyldithiocarbamate	20941-65-5
Dinonyldiphenylamine	24925-59-5
Diphenyl cresyl phosphate	26444-49-5
4-Isopropyl phenyl isocyanate	31027-31-3
Styrenated diphenylamine	68442-68-2

Dicyclopentadiene

1. Summary and assessment

Following oral administration, dicyclopentadiene (DCPD) is eliminated mainly in the urine and faeces (half-life: 18 hours in the rat and in the mouse, 17 hours in the dog).

The oral LD_{50} value for rats ranges betweend 350 and 800 mg/kg. On contact with the skin or mucosa (eyes and respiratory tract), DCPD causes irritation, increasing in degree with concentration. In humans, irritative effects on the upper respiratory tracts and the eyes must be expected at about 10 ppm, and in sensitive individuals even at 1 ppm.

DCPD is not considered to be sensitizing (the experimental results are difficult to interpret).

No toxic effects are observed following subchronic oral administration to rat, mouse, dog or weasel in doses of up to 200 mg/kg. Subchronic inhalation of DCPD (35 and 74 ppm) causes kidney lesions and loss of body weight in rats, especially in male animals ("no effect level" approx. 20 ppm).

DCPD manifests no mutagenic properties in the Salmonella microsome test or when tested on Saccharomyces cerevisiae.

According to one brief report that DCPD has no carcinogenic potential when injected intramuscularly into rats; there are, however, no detailed reports available as yet.

No evidence of toxic effects of DCPD on reproduction is found in weasels.

In the USA the TLV is given as 5 ppm (approx. 27 mg/m^3; TLV, 1980) and in Russia as 0.185 ppm (approx. 1 mg/m^3; Korbakova, 1964; Shashkina, 1965).

2. Name of substance

2.1	Usual name	Dicyclopentadiene
2.2	IUPAC name	Tricyclo[5.2.1.0^{2,6}]deca-3,8-diene
2.3	CAS-No.	77-73-6

Dicyclopentadiene

3. Synonyms

Common and trade names   CP-dimer
                                DCPD

4. Structural and molecular formulae

4.1 Structural formula

4.2 Molecular formula $C_{10}H_{12}$

5. Physical and chemical properties

5.1 Molcecular mass   132.20 g/mol
5.2 Melting point   32.5–33.6° C
5.3 Boiling point   166.6–172.8° C (at 1013 hPa)
5.4 Density   0.9816 g/cm^3 (20° C)
5.5 Vapour pressure   1.86 hPa (20° C)
                                4.0 hPa (20° C)
5.6 Solubility in water   40 ppm
5.7 Solubility in organic solvents   soluble in ethanol, soluble in ether
5.8 Solubility in fat   no information available
5.9 pH-value   no information available
5.10 Conversion factor   1 ppm $\stackrel{\wedge}{=}$ 5.4 mg/m^3
                                  (at 25° C and 1013 hPa)
                                1 mg/m^3 $\stackrel{\wedge}{=}$ 0.185 ppm
5.11 Saturation in air   1842 ppm
                                  (at 20° C and 1013 hPa)
5.12 Odour   similar to camphor umpleasant (Scherb, 1975; Kinkead et al., 1971; Rosenblatt et al., 1965; Patty, 1963)

6. Uses

Starting material for the manufacture of alkyd resins, synthetic rubber, terpenes, alkaloids, insecticides, hydraulic fluids, plasticizers and paints (Scherb, 1975).

7. Results of experiments

7.1 Toxicokinetics and metabolism

Three different studies were performed on the animal species listed below in order to determine the kinetics of DCPD (Hart, 1976).
- 24 male Swiss-Webster mice, 20–30 g, 40 mg ^{14}C-DCPD/kg p.o.
- 12 male Sprague-Dawley rats, 180–280 g, 100 mg ^{14}C-DCPD/kg p.o.
- 5 male beagle dogs, 7.6–8.9 kg, 100 mg ^{14}C-DCPD/kg p.o.

3 mice were sacrificed 15 minutes and 1, 2, 4, 6, 24, 48 and 72 hours after administration of the substance, and 2 rats were sacrificed after 2, 4, 6, 24, 48 and 72 hours. In the case of the dogs, blood samples were taken after 30 minutes and after 1, 2, 4, 6, 10 and 24 hours. Maximum plasma concentrations of 11.36 and 39.95 µg ^{14}C-DCPD/ml plasma were found after 2 hours in the mice and dogs, whilst in rats the maximum concentration of 23.28 µg/ml was determined 6 hours after administration. The authors established a 2-phase elimination courve for all 3 animals species. In the rapid phase, the half-life was 4 hours for mice and 10 hours for the dogs. Insufficient values were available for the rats for an exact calculation to be made. The authors measured half-lives of 18 hours (mice), 27 hours (dogs) and 18 hours (rats) during the slow phase. In all 3 animal species the radioactivity was distributed through-out the organism within 1 to 2 hours. Maximum levels were found in the bladder (empty), bile and body fat of the mice, in the bile and gall bladder (empty) of the dogs and in the fat, adrenals and bladder (empty) of the rats. All 3 species eliminated a large portion of the radioactivity within 24 hours:

	urine (%)	faeces (%)
mouse	75	10
rat	53.7	15
dog	64.5	4

Two metabolites were found in the urine of mice and dogs, and three in the urine of rats, but these could not be identified by thin-layer chromatography (Hart, 1976).

A further toxicokinetic study was performed on a Jersey cow. The animal received 10 mg DCPD/kg/day p.o. for 5 days, with a dose of 4.0×10^8 dpm ^{14}C-marked DCPD (equivalent to 10 mg DCPD/g) on day 6. A maximum concentration of 290 dpm/g (equivalent to 2.1 µg DCPD) was found in the blood after 2 hours. Within 24 hours the cow had eliminated 80.1% of the radioactivity via the urine, 3.7% via the faeces and 0.07% via the milk. 96 hours after administration, no radioactivity could be found in the brain, fat, gall bladder, heart, kidneys, liver, muscle, ovaries, lungs, adrenals, skin, spleen, bladder and udder (detection limit in blood, 20 dpm/g (equivalent to 0.15 µg DCPD/g); no data were available for tissue). The authors of this study were satisfied that the exposure of cattle to low levels of DCPD would not result in any appreciable contamination of the meat or the milk (Ivie and Oehler, 1980).

7.2 Acute and subacute toxicity

The LD_{50} for rats lay between 347 and 830 mg/kg following oral administration and between 200 and 300 mg/kg following intraperitoneal administration; for mice it lay between 190 and 250 mg/kg following oral administration and at 200 mg/kg following intraperitoneal administration (Table 1). Ataxia, reduced motility, exhaustion, tremor and cramps were common features of acute intoxication in mice and rats. Mice and rats which died prematurely during the subsequent observation period manifested hyperaemia of the lungs. In addition to this, mice showed an accumulation of yellow fluid in the stomach and small intestine, distension of the bladder with pinkish-orange fluid, and black discolouration of portions of the liver and spleen. Hart indicates that some of the symptoms observed could be a consequence of autolysis. Dissection at the end of the 14-day observation period revealed no microscopic changes in either mice or rats (Kinkead et al., 1971; Hart, 1976).

In weasels the symptoms of acute intoxication were marked by hyperactivity, shrill cries, dyspnoea, diarrhoe, opisthotonus, cramps, vomiting and paresis of the hind limbs. Recovery occurred within 1 to 1.5 hours. A dose of more than 960 mg/kg body weight p.o. proved lethal. On microscopic examination, no pathological changes were seen in the animals which died or were sacrificed at the end of the test (Aulerich et al., 1979).

A single oral dose of 250, 500, 1000, and 2000 mg DCPD/kg in 8 to 10 weeks old calves (4 animals per group) resulted in toxic reactions at all dosages. The dose-dependent reactions included ataxia and increased salivation from 250 mg/kg body weight and tonic-clonic cramps from 500 mg/kg. The two higher doses lay in the lethal range: one of the 4 calves died after 3 days at 1000 mg/kg body weight, whilst at 2000 mg/kg all 4 animals died within 7 days. At a dose of 500 mg/kg or more, DCPD produced a dose-dependent increase in the activity of creatine-phosphokinase (CPK), aspartate-aminotransferase (AST) and alanine-aminotransferase (ALT), together with an increase in leucocytes, RBC, haemoglobin and haemotocrit values. In the surviving calves these parameters returned to normal within 14 days. Microscopic examination of the brain, spinal chord, lungs, heart, liver, kidneys, bladder, spleen, mesenterial lymph nodes, stomach, small and large intestine, epididymis or ovaries, oviduct, uterus, thyroid, pancreas and adrenals showed no histopathological changes. No comparable data from control animals were available in this study (Cysewski et al., 1981).

In an acute inhalation test ("range-finding toxicity data") all rats (6 animals/test) died within 30 minutes or 2 hours of exposure to saturated DCPD vapour (more accurate data lacking; Smyth et al., 1954).

In a second test, the inhalation of 500 ppm DCPD over 4 hours resulted in the death of 6 rats, whilst 1000 ppm caused death of all the exposed rats (Smith et al., 1962).

In a further study performed by the same investigators, 4 out of 6 rats died within 14 days of a 4-hour exposure at 2000 ppm. At a concentration of 2500 ppm for 1 hour, 1 out of 4 rats died, whilst at 1000 ppm over 4 hours, 4/4 rats died (Gage, 1970). For individual data see Table 2.

In a range finding test in rats the 4-hour LC_{50} was 660 ppm (Kinkead et al., 1971).

In more accurate studies by the same investigators (see Table 3; concentration determined using a gas chromatograph with a flame ionisation detector), the 4-hour LC_{50} values varied on the one hand between 145.5 (117.5–180.2) ppm in mice, 359.4 (290–445.1) ppm in male rats, and 385.2 (311.1–477.1) ppm in female rats and, on the other hand, 770.5 (595.6–996.6) ppm in guinea pigs and 771.0

(505.2–1177.0) ppm in rabbits. The inhalation of 773 ppm over 1 hour was fatal to 1 female dog, whilst each of the concentrations of 458, 272 and about 68 ppm over 4 hours were survived by 1 female dog. Dose-dependent eye irritation and incoordination were observed in all species. Cramps occurred when the animals died. The individual data are given in Table 3 (Kinkead et al., 1971).

No reactions were observed in beagles following the administration of 40, 125 and 375 ppm of DCPD in the diet for 14 days. The corresponding doses were roughly 5–12, 15–32 and 49–64 mg of DCPD/kg/day. One dog of each sex was used per dose. Behaviour, food intake, body-weight gain, haemotological, biochemical and urine analyses, microscopic sectioning and histology were used to evaluate the results in this study. It should be pointed out that even with the highest doses tested there was no effect on body weight. An investigation performed in accordance with current testing techniques would have included a maximum tolerated dose (MTD) (Hart, 1976).

In a test on eight-months-old weasels DCPD was administered in the diet for 21 days at levels of 1, 10, 100, 1000 and 10000 ppm. This is equivalent to about 0.17, 1.9, 17.7, 178 and 754 mg/kg/day. 5 animals of each sex were used per dose. At levels of up to 1000 ppm, there was no effect on haematocrit values and blood counts. No changes in comparision with the controls could be observed either macroscopically or microscopically (brain, heart, lungs, kidneys, spleen, and liver). In the first week of the test, weight gain was less than in the controls for both male and female animals, but this effect was no longer seen in the 2nd and 3rd week. A mortality of 68% was reported for the maximum dose (10000 ppm). At the same time there was a significant reduction in food consumption (in comparison with the controls: about 75% less) and in body weight (in comparison with the controls: about 66% less). The animals surviving at the end of the test (21 days) showed a significantly reduced haemotocrit value, a reduced neutrophil count and marked loss of body fat. The authors considered it possible that there was a connection between reduced food intake and mortality, as well as the other side effects observed (Aulerich et al., 1979).

6 male and 6 female Harlan-Wistar rats were exposed to 72, 146 and 332 ppm DCPD vapour in inhaled air (purity 98.3%) for 7 hours/day on 5 days/week over a total of 10 days. No adverse reaction or macroscopically visible changes in the internal organs were observed following exposure at either of the low concentrations

(72, 164 ppm) in comparison with control animals. At the highest concentration (332 ppm) the males developed cramps during the 2nd inhalation stage and thereafter. All males and females in this group died after 1–4 days. Macroscopic investigation showed that 332 ppm caused bleeding in the lungs and intestine in all cases, together with bleeding in the thymus of the females (Kinkead et al., 1971).

In a further subacute inhalation test (2 male and 2 female Alderly-Park rats per concentration) 1 out of 4 rats died after the daily inhalation of 250 ppm over 6 hours repeated 10 times. The surviving animals suffered loss of weight, nasal irritation, dyspnoea, lethargy, tremor and increased sensitivity. The blood parameters investigated (Hb, erythrocytes, leucocytes, total blood count, thrombocytes, clotting time, urea, sodium and potassium) showed no pathological changes. A level of 100 ppm (15 exposures of 6 hours each) was tolerated (see also Table 2) (Gage, 1970).

The effect of DCPD vapour (47, 72, and 146 ppm inhaled air) on albino mice was tested in the same way using the same number of animals as in the case of the rats. Whereas the low concentrations of DCPD (47 ppm) were tolerated without reaction, 5 of each of the 6 male and female animals died within 10 days following exposure to 72 ppm. The body-weight gain of the 2 surviving mice was normal. At the highest concentration of DCPD (146 ppm), all of the animals suffered cramps and died during the first stage of the test. On macroscopic examination, no pathological changes could be seen in either the mice that died or those that were sacrificed at the end of the test (Kinkead et al., 1971).

In another study beagles were exposed to DCPD vapours at concentrations of 20, 47 and 72 ppm inhaled air (same test design as for rats and mice). One animal was exposed per dose. In this study 20 ppm resulted in mild diarrhoea. At 47 ppm, diarrhoea, increased salivation and incoordination of the hind limbs occurred. At the higher concentration (72 ppm) these symptoms were not observed. Body-weight gain was normal for all animals. Marcroscopic investigation showed no treatment-related changes (Kinkead et al., 1971).

7.3 Skin and mucous membrane effects

0.5 ml of undiluted DCPD produced marked irritation of the conjunctiva in rabbits (Smyth et al., 1954, 1962).

In another test using Draize method, the sensitivity of the conjunctiva was tested in 9 rabbits (New Zealand White). In this case, irritation of the conjunctiva occurred in 7 of the exposed animals following the instillation of 0.1 ml of DCPD (no information on concentration). Even thorough rinsing of the eyes 2 or 4 seconds after administration of the substance effected no improvement. The conjunctiva had however returned to normal in all cases 3 days following application (Hart, 1976).

The application of 0.01 ml of undiluted DCPD to the shaved skin of the back (uncovered) (Smyth et al., 1954) or skin of the abdomen (uncovered) (Smith et al., 1962) of 5 albino rabbits caused moderately severe irritation of the skin.

Following the application of 2 g DCPD/kg to the shaved intact skin or abraded skin (occluded over 24 hours) of 2 rabbits in each case, mild erythema occurred in 3 animals (accurate data lacking). Complete healing occurred after 14 days. Oedema, crusting and systemic toxicity were not observed (Hart, 1976).

Dermal toxicity was tested in rabbits with the application of 4.38, 4.99 and 6.60 g/kg (occluded, 24 hours) (Kinkead et al., 1971; Smyth et al., 1954, 1962). The authors provided no data on intoxication symptoms due to absorption.

7.4 Sensitization

DCPD was tested for sensitizing properties on guinea-pigs. 8 guinea-pigs received 0.1% DCPD (w/v) intracutaneously on 10 occasions (3 times per week). The volume administered was 0.05 ml in the first instance, subsequently 0.1 ml. No skin reactions were observed. Two weeks after the last administration the animals were injected again with 0.05 ml of a 0.1% solution of DCPD as a challenge. Only slight erythema was observed 24 and 48 hours later. Oedema was not observed (number of reacting animals not given, evaluation was the local reaction on the basis of the Draize classification). The authors regarded the results as negative. 2,4-Dinitro-1-chlorobenzene, which was administered to 4 control guinea-pigs, caused a marked skin reaction (Hart, 1976).

7.5 Subchronic-chronic toxicity

ARS-Sprague-Dawley rats received 80, 250 and 750 ppm of DCPD (purity 98–99%) in the diet (corresponding to about 5–22, 16–62 and 44–198 mg DCPD/kg/day for 90 days. 30 animals of each sex were used per dose. No effect of DCPD could be detected for any of the 3 levels tested (80, 250 and 750 ppm) (Hart, 1976).

In another study DCPD was administered to ICR-Swiss-Albino mice at levels of 28, 91 and 273 ppm in the diet (corresponding to about 4–10, 21–31 and 37–88 mg/kg/day) for 90 days. 32 males and 32 females formed each dose group. Again, no reactions were observed in this test.

In both studies the test parameters were behaviour, food intake, body-weight gain, haemotological, biochemical and urine analyses and post-mortem investigations including the histology of 31 organs. It should be mentioned that even in the high dose group no inhibition of body-weight gain was observed in comparison with control animals. A test in accordance with present-day testing techniques would have included a maximum tolerated dose (MTD). Data on preliminary tests for determination of the applied doses are lacking (Hart, 1976).

12 male and 12 female Harlan-Wistar rats were exposed to DCPD vapours at concentrations of 19.7, 35.2 and 73.8 ppm in inhaled air (purity 98.3%) for 7 hours/day, 5 days/week for 89 days of treatment (about 18 months). On test days 4, 13, 31, 55, 75 and 89, clinical investigations were performed for possible symptoms of toxicity and body weight was determined. Livers and kidneys were weighed at the end of the test. Following an extensive macroscopic examination, the heart, liver, kidneys, lungs, spleen, adrenals, trachea, prostate, testes, colon and mesenterium of each animal were examined microscopically. At all 3 concentrations of DCPD (19.7, 35.2 and 73.8 ppm) there was a significant increase in the absolute and relative kidney weight of the males. However, no concentration-dependence could be found. Although no further changes were observed at 19.7 ppm, body-weight gain was affected at medium concentration (35.2 ppm). Microscopic examination revealed effects on the kidneys, particularly in the males, with round cell infiltration, potassium and protein deposits, together with widening and degeneration of the kidney tubules and nephrolithiasis. These pathological changes were clearly observed at the highest DCPD concentration (73.8 ppm). The body-weight curve was markedly flattened, with the values significantly lower than in the control animals on the 4th day of treatment. In addition to this, there was a significant increase in the relative liver weight in male rats, without any histological changes. The other organs examined microscopically showed no effect of the substance at any of the 3 DCPD concentrations. According to the authors' data the no-effect-level lies between 19.7 and 35.2 ppm (Kinkead et al., 1971).

In another study, the same authors administered 8.9, 23.5 and 32.4 ppm DCPD vapour in inhaled air to male beagles for 7 hours/day, 5 days/week for 89 days of treatment. Haematological and biochemical investigations (haematocrit, white blood cell count, blood urea, SAST, SALT, acid and alkaline phosphatase) were carried out 6 days before the start of the test and on days 20, 37, 65 and 85 of treatment. Urine was tested 5 days before the start of treatment and on days 21, 38, 68 and 87. In addition to this, the animals were regularly observed for signs of intoxication. Liver and kidney weights were determined at the end of the test. No treatment-related changes occurred following the administration of 8.9 ppm. For the middle and high DCPD concentration (23.5 and 32.4 ppm), increase in the activity of acid and alkaline phosphatase and of serum aspartate aminotransferase (SAST) were observed on some test days, to- gether with increased urea values and a reduced neutrophil count. These observed changes however only occurred in isolation and were in no way concentration-dependent. The authors considered them to be without physiological significance. All the other haematological and biochemical parameters investigated remained unaffected. This also applied to the urine analysis. Microscopically no pathological changes were found in lungs, liver, kidneys, heart, spleen, adrenals, thyroids, parathyroid, oesophagus, diaphragm, lymph nodes, gall bladder, salivary glands, tongue, stomach, duodenum, pancreas, ileum, jejunum, colon, bladder, prostate, testicles, epididymis, brain, hypophyses, skin and eyes. As slight pathological changes in biochemical parameters occurred on some test days at 23.5 ppm, the authors considered that the no-effect-level lay between 8.9 and 23.5 ppm (Kinkead et al., 1971).

7.6 Genotoxicity

DCPD was tested for mutagenic properties on eucaryontes (Sacharomyces cerevisiae D3, mitotic recombination test) and procaryontes (Salmonella typhimurium TA 98, TA 100, TA 1535, TA 1537, and TA 1538). With Saccharoymces cerevisiae D3, concentrations of 0.01–0.09% were used in the first test and 0.01–0.017% in a repeat test. In the Salmonella/microsome test, a concentration of 10–5000 µg/plate was used, with a concentration of 10–500 µg/plate in a repeat test. The repeat tests were performed because toxic effects occurred at the higher levels. All the tests were per- formed with and without metabolic activation. The results of the studies which were performed in accordance with the current state of testing techniques provided no indication of mutagenic properties (Simmon and Kauhanen, 1978).

The lack of any mutagenic effects by DCPD in Salmonella typhimurium has been confirmed in another investigation (NTP, 1982).

7.7 Carcinogenicity

The question of the possible carcinogenic properties of DCPD was investigated in 1975 by the Institute of Chemical Biology, San Francisco University, under contract to the National Cancer Institute. No indications of carcinogenic properties of DCPD were found following the intramuscular administration of DCPD to rats under the selected test condition, not described in detail (detailed data currently not available; Rosenblatt et al., 1975).

7.8 Reproductive toxicity

3-month-old weasels received DCPD at levels of 100, 200, 400, and 800 ppm (equivalent to 23.6, 42.4, 85.0 and 169.9 mg/kg/day) in the diet for 12 months. One male and 3 females were kept together in one cage. Four such groups were used per dose. The investigation was timed in such a way that mating and birth and rearing of the young took place during the last third of the test. The test parameters were as follows: mortality, weight gain (determined every 2 weeks), food intake (determined every 2 weeks), blood investigation (haematocrit and total blood count every 3 months), duration of pregnancy, litter size, weight of the young animals, sex ration among the young and mortality. After the test the mothers were dissected and the following organs were weighed: brain, liver, kidneys, spleen, gonads, lungs, heart and adrenals. In addition to this, a further 4 organs were examined histologically. Mortality, body-weight gain, food intake, blood count and reproduction (reproductive capacity of the males, lenght of pregnancy, litter size, litter weight and sex ration of the young) were not adversely affected. Four weeks after discontinuation of treatment, the young animals from the 200, 400, and 800 ppm groups had a significantly lower body weight. The mortality of the young animals was not significantly affected up to this time. Macroscopic and microscopic examination of the organs at the end of the test revealed no reliable treatment-related changes. However, the weight of the spleen was significantly reduced in the animals given 400 ppm, and the weight of the testicles was significantly reduced in the animals given 800 ppm. There were no pathological changes in the testicles. According to the authors, this was likely to be a random finding (Aulerich et al., 1979).

7.9 Effects on the immune system

No information available.

7.10 Other effects

In order to test a possible effect of DCPD on the hexobarbital sleeping time (enzyme induction) of rats (Charles-River-COBS-CD(SD)BR), 2 groups (of 10 animals of each sex) were treated with 750 ppm in the died. On the 5th day of DCPD-administration, one group received an intraperitoneal injection of 100 mg hexobarbital/kg body weight. There was no change in sleeping time in comparison with the controls (Hart, 1976).

8. Experience in humans

In one test two volunteeres were exposed to 1 ppm DCPD in an inhalation chamber for 30 minutes. One test subject suffered eye and throat irritation after 7 minutes, while the other person reported a reduced sense of smell (Kinkead et al., 1971).

The same 2 subjects were exposed to a concentration of DCPD of 5.5 ppm in inhaled air after a break of 30 minutes. They reported no change in the sense of smell. One suffered irritation of the eyes after 10 minutes, whilst the second reported that he could still taste DCPD 1 hour later (Kinkead et al., 1971).

During a 5-month observation period, it was found that an accidental exposure of workers to DCPD resulted in occasional transient headaches during the first 2 months. This reaction no longer occurred during the following 3 months. The authors see this as an indication of a slowly increasing tolerance threshold (Kinkead et al., 1971).

Irritation of the mucosa is to be expected in man at DCPD concentrations above 10 ppm (other data lacking in the reference; TLV, 1980).

The odour threshold for DCPD in man has been given as 0.003 ppm or 0.05 ppm. However, the first value applies to a test with only 3 subjects, who were selected at random with no prior testing of their sense of smell. Precise information of the derivation of the latter value is lacking (Cherb, 1975; Kinkead et al., 1971).

Table 1. Acute toxicity of dicyclopentadiene

Animal species	Animals/dose	Sex	Route of administration	Observation period (days)	LD_{50}	References
Rat (Carworth-Wistar)	–	Male	Oral	14	0.41 ml/kg ($\hat{=}$ 400 mg/kg)	Smyth et al., 1962
Rat	–	Male	Oral	–	435 mg/kg	Rosenblatt et al., 1975
Rat (Sprague-Dawley)	10	Male	Oral	14	520 mg/kg	Hart, 1976
Rat (Carworth-Wistar)	–	Male	Oral	14	820 mg/kg	Smyth et al., 1954
Rat (Sprague-Dawley)	10	Female	Oral	14	378 mg/kg	Rosenblatt et al., 1976
Rat	–	Female	Oral	–	396 mg/kg	Hart, 1976
Rat (Albino)	–	–	Oral	14	0.353 ml/kg ($\hat{=}$ 347 mg/kg)	Kinkead et al., 1971
Rat	–	–	i.p.	14	0.31 ml/kg ($\hat{=}$ 300 mg/kg)	Kinkead et al., 1971
Mouse (Swiss-Webster)	10	Male	Oral	14	190 mg/kg	Hart, 1976
Mouse (Swiss-Webster)	10	Female	Oral	14	250 mg/kg	Hart, 1976

Table 1. (Continued)

Animal species	Animals/dose	Sex	Route of administration	Observation period (days)	LD$_{50}$	References
Rabbit	4	Male	dermal (24 hours)	14	4.46 ml/kg (= 4.38 g/kg)	Smith et al., 1962
Rabbit	4	Male	dermal (24 hours)	14	6.72 ml/kg (= 6.60 g/kg)	Smyth et al., 1962
Rabbit	–	–	dermal	–	5.08 ml/kg (= 4.99 g/kg)	Kinkead et al., 1971
Weasel	2–4	Male	oral	14	> 1 000 mg/kg	Aulerich et al., 1979
Weasel	1	Female	i.p.	–	at 960 mg/kg 1/1 dead in few minutes	Aulerich et al., 1979
Calf	4	Male + female	oral	14	ca. 1 000 mg/kg	Cysewski et al., 1981

– No information available

Table 2. Acute inhalation toxicity of dicyclopentadiene (Gage, 1970)

Animal species	Animals/dose	Dose (ppm)	Exposure period (hours)	Observation	Comments
Rat (Alderly Park)	2 male 2 female	2 500	1	1/4 dead[a]	Eye and nose irritation, dyspnoea, loss of consciousness; dissection: congestion in the lungs, liver and kidney
Rat (Alderly Park)	2 male 2 female	1 000	4	4/4 dead[a]	Eye and nose irritation, dyspnoea, loss of muscular coordination, tremor, increased sensitivity; dissection: congestion in the lungs, liver and kidney
Rat (Alderly Park)	2 male 2 female	250	10x6	1/4 dead[a]	Loss of weight, nasal irritation, dyspnoea, lethargy, tremor, increased sensitivity, blood values normal; dissection: no abnormalities
Rat (Alderly Park)	4 male 4 female	100	15x6	no deaths	No signs of intoxication; dissection: no abnormalities

[a] No sex-related information available

Table 3. Acute inhalation toxicity of dicyclopentadiene (Kinkead et al., 1971)

Animal species	Sex	Weight (g)	Animals/dose	Exposure period (hours)	LC_{50} and range (ppm)	Comments
Mouse	Male	31–41	6	4	145.5 (117.5–180.2)	272 ppm: tonic cramps in one mouse after 75 minutes, all animals died during the night; 110 ppm: no signs of intoxication, 1 late death
Rat	Male	105–214	6	4	359.4 (290.2–445.1)	272 ppm: irritation in the extremities within 60 minutes, 1 animal died
Rat	Female	100–176	6	4	385.2 (311.1–477.1)	272 ppm: irritation in the extremities within 60 minutes, 1 animal died
Guinea pig	Male	655–917	6	4	770.5 (595.6–996.6)	458 ppm: very slight loss of coordination, no deaths
Rabbit	Male	1 912–2 568	4	4	771.0 (505.2–1177.0)	458 ppm: incoordination within 180 minutes, no death

Species	Sex	Dose			Observations
Dog (Beagle)	Female	7 100	1	1	–	773 ppm: irritation of the eye, nose and extremites followed by tear secretion within 30 minutes, the dog survived
Dog (Beagle)	Female	10 800	1	4	–	458 ppm: irritation of the eye, nose and extremities followed by tear secretion within 50 minutes, the dog survived
Dog (Beagle)	Female	7 700	1	4	–	272 ppm: tremor within 180 minutes, the dog survived
Dog (Beagle)	Female	7 600	1	4	–	ca. 68 ppm: frequent slight irritation immediately after administration, the dog survived

– No information available

References

Aulerich, R.J., Coleman, T.H., Polin, D., Ringer, R.K., Howell, K.S., Jones, R.E., Kavanagh, T.J.
Toxicology study of diisopropyl methylphosphonate and dicyclopentadiene in Mallard Ducks, Bobwhite Quail and Mink
Environmental Protection Research Division, U.S. Army Medical Bioengineering Research and Development Laboratory
Report AD-AO87 257 (1979)

Cysewski, S.J., Palmer, J.S., Crookshank, H.R., Steel, E.G.
Toxicological evaluation of diisopropyl methylphosphonate and dicyclopendadiene in cattle
Arch. Environ. Contam. Toxicol., 10, 605–615 (1981)
auch veröffentlicht als Report AD-AO93 673, Autoren: Palmer, J.S., Cysewski, S.J. (1979)

Gage, J.C.
The subacute inhalation toxicity of 109 industrial chemicals
Br. J. Ind. Med., 27, 1–18 (1970)

Hart, E.R.
Mammalian toxicological evaluation of DIMP and DCPD
Environmental Protection Research Division, U.S. Army Medical Bioengineering
Research and Development Laboratory
Report AD-AO58 323 (1976)

Ivie, G.W., Oehler, D.D.
Fate of dicyclopentadiene in a lactating cow
Bull. Environ. Contam. Toxicol., 24, 662–670 (1980)

Kinkead, E.R., Pozzani, U.C., Geary, D.L., Carpenter, C.P.
The mammalian toxicity of dicyclopentadiene
Toxicol. Appl. Pharmacol., 20, 552–561 (1971)

Korbakova, A.I.
Standard levels of new industrial chemicals in the air of work premises
Vestn. Akad. Med. Nauk SSSR, 19, 17–23 (1964), Kurzreferat in Chem. Abstr., 61, 16694c (1964)

NTP Technical Bulletin, 7, 5–7 (1982)
U.S. Department of Health and Human Services

Patty, F.A. (ed.)
Industrial Hygiene and Toxicology
2nd ed., 1209–1217, Wiley (Interscience), New York (1963)

Rosenblatt, D.H., Miller, T.A., Dacre, J.C., Muul, I., Cogley, D.R.
Problem Definition Studies on Potential Environmental Pollutants
II. Physical, Chemical, Toxicological and Biological Properties of 16 Substances
U.S. Army Medical Bioengineering Research and Development Laboratory
Technical report AD-AO30 428, J1–8 (1975)

Scherb, H.
Cyclopentene und -pentadiene
Ullmanns Encyclopaedie der technischen Chemie, 4. Auflage, Band 9, 699–704 (1975)
Verlag Chemie, Weinheim

Shashkina, L.F.
The maximum permissible concentration of cyclopentadiene and dicyclopentadiene in the atmosphere of industrial premises
Gig. Tr. Prof. Zabol., 9, 13–19 (1965), Kurzreferat in Chem. Abstr., 64, 20509c (1965)

Simmon, V.F., Kauhanen, K.
In vitro microbiological mutagenicity assays of dicyclopentadiene
SRI Project LSU-5612 (1978)

Smyth, H.F., Carpenter, C.P., Weil, C.S., Pozzani, U.C.
Range-Finding Toxicity Data: List V
Arch. Ind. Occup. Med., 10, 61–68 (1954)

Smyth, H.F., Carpenter, C.P., Weil, C.S., Pozzani, U.C., Striegel, J.A.
Range-Finding Toxicity Data: List VI
Am. Ind. Hyg. Assoc. J., 23, 95–107 (1962)

TLV, Fourth Edition
American Conference of Governmental Industrial Hygienists Inc., Cincinnati, Ohio (1980)

2-Methylpropanol-1

1. Summary and assessment

2-Methylpropanol-1 is rapidly metabolised to isobutyraldehyde and isobutyric acid; small quantities of 2-methylpropanol-1, isobutyraldehyde and isobutyric acid are found in the urine, chiefly combined with glucuronic acid. In contrast to ethanol, 2-methylpropanol-1 is distributed not only in water-containing but also in water-free tissues.

The acute oral toxicity of 2-methylpropanol-1 (isobutylalcohol) is low (oral LD_{50} for the rat is 2.5 to 3.1 g/kg). In acute trials using various modes of administration, the following toxic effects have been observed: apathy, narcosis, irritation of the mucous membranes, liver and kidney lesions. 2-Methylpropanol-1 has a weak to moderate irritative effect on both the intact and the scarified skin (administered occlusively), and leads to mild to severe irritation on the mucosa of the eye.

Administration in the drinking water over a 4-month period results in gastrointestinal disorders (dyspepsia, mucosal bleeding). Reversible fatty changes in the liver are described following repeated inhalation of narcotic concentrations.

The positive finding of a mutagenic effect of 2-methylpropanol-1, reported in one study, cannot be assessed since suspected mutations were found only in E. coli which had been pre-treated with diluted (2.5%) 2-methylpropanol-1 (survival rate 1–10%).

Injecting 2-methylpropanol-1 into the yolk-sac of fertilized eggs has a toxic effect. There are no reports of deformities in the hatched chicks. Studies of possible embryotoxic effects in mammals have not been undertaken.

The data available in the literature on a possible carcinogenic effect of 2-methylpropanol-1 following oral and subcutaneous administration to rats cannot be assessed since the reported findings do not permit either a qualitative or quantitative interpretation.

In humans, reversible eye lesions can occur following exposure to 2-methylpropanol-1. Following oral administration of ethanol and 2-methylpropanol-1 (the latter is present in alcoholic drinks) there is a deterioration in performance during the phase of acute alcoholization, correlated with the ethanol concentration in the blood. In the post-alcoholization phase, the frequency of errors associated with the level of 2-methylpropanol-1 increases noticeably. The odour threshold in the air is given as 1.6 ppm.

The MAK value (maximum workplace concentration) is 100 ppm (300 mg/m^3; Deutsche Forschungsgemeinschaft, 1986).

A 90-day trial has been commissioned by the BG Chemie to test for a possible chronic effect, and, if indicated, a study of carcinogenicity will follow. Furthermore, a teratogenicity study is also being carried out on behalf of the BG Chemie.

2. Name of substance

2.1	Usual name	2-methylpropanol-1
2.2	IUPAC name	2-methylpropane-1-ol
2.3	CAS No.	78-83-1

3. Synonyms

Common and trade names
isobutyl alcohol
isobutanol
1-hydroxymethylpropane
2-methylpropylalcohol
2-methylpropan-1-ol

4. Structural and molecular formulae

4.1 Structural formula

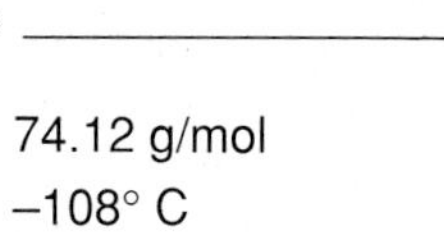

$$CH_3-CH(CH_3)-CH_2-OH$$

4.2 Molecular formula $C_4H_{10}O$

5. Physical and chemical properties

5.1	Molecular mass	74.12 g/mol
5.2	Melting point	−108° C
5.3	Boiling point	108° C
5.4	Density	0.806 g/cm^3 (at 20° C)
5.5	Vapour pressure	11.7 hPa (at 20° C)
5.6	Solubility in water	95 g/l

5.7	Solubility in organic solvents	miscible with alcohols and ether
5.8	Solubility in fat	no information available
5.9	pH-value	no information available
5.10	Conversion factors	1 ppm $\hat{=}$ 3.03 mg/m^3 1 mg/m^3 $\hat{=}$ 0.33 ppm (Windholz, 1983; Kühn-Birett, 1980).

6. Uses

Solvent for oils, fats, waxes, natural resins and plastics; the paint industry is the main user; extraction agent; esters of phthalic acid, adipic acid and other dicarboxylic acids and phosphoric acid are used as plasticizers (Ullmann, 1975).

7. Results of experiments

7.1 Toxicokinetics and metabolism

5 ml of liver homogenate (rat; 0.25 g liver/ml) was incubated with 80 µmol of 2-methylpropanol-1 for 30 minutes. In this time 0.21 mmol/g liver was metabolised; the equivalent amount for ethanol was 0.16 mmol/g liver (Hedlund and Kiessling, 1969).

In the liver perfusion preparation in situ, 5.3 µmol of 2-methylpropanol-1 in 200 ml of perfusion fluid (heparinized human blood diluted 1:1 with physiological NaCl solution) was perfused for 30 minutes (2 ml/minute); under these test conditions, 0.065 mmol/g liver was metabolized. The rate of metabolism was higher than with ethanol (0.05 mmol/g liver; Hedlund and Kiessling, 1969).

Kühnholz et al., (1984) determined the solubility of aliphatic alcohols in the body tissues of rabbit and man in vitro. 5 ml of physiological NaCl with a concentration of 10 mg/l of the alcohol were incubated with pulverized body tissue for a minimum of 45 minutes at 37° C. Tests were conducted on muscle, brain, lungs, kidneys, fatty tissue, liver, spleen and blood. Solubility was similar in the various types of tissue. There were no basic differences between rabbit and human tissues. In contrast to short-chain primary alcohols (methanol, ethanol), 2-methylpropanol-1 was distributed not only in aqueous tissues, but also in anhydrous tissues. It was deduced from this fatty tissue (water content approx. 10–20%) acts as a depot for the longer-chain alcohols.

Gaillard and Derache (1966) found that 2-methylpropanol-1 (4 g/kg in 50% aqueous solution, orally) had no influence on the lipid content of the liver and blood in the rat (8 Wistar rats, 220 g). 17 hours after administration of 2-methylpropanol-1, the lipid content of the liver and the concentration of triglycerides, cholesterol and phospholipids in the blood were unchanged. Gaillard and Derache (1965) also investigated the rate of elimination of 2-methylpropanol-1 from the blood after oral administration of 2.0 g/kg in 20% aqueous emulsion (tests on 5 Wistar rats, mean weight 300 g, sex not stated). For this purpose, blood was taken from the caudal vein and analysed at 0, 15, 30, 60, 90, 120, 240 and 480 minutes after administration. The urine was also collected during this period and the amount of 2-methylpropanol-1 excreted was determined. The authors found 4 mg/100 ml in the blood after 15 minutes. A peak concentration of 24 mg/100 ml occurred after 90 minutes and after 8 hours 2 mg/100 ml could still be detected. In the urine, 52 mg/100 ml of the oral dose of 2-methylpropanol-1 was found after 90 minutes, and 27 mg/100 ml was still detectable after 8 hours.

7.2 Acute and subacute toxicity

The intraperitonal LD_{50} of 2-methylpropanol-1 in male ICR mice (7 days follow-up observation) was put at 312 mg/kg (Dillingham et al., 1973).

Other authors found an intraperitoneal LD_{50} in male Swiss-Webster mice (20–25 g) of 544 mg/kg (384–735 mg/kg), with a follow-up period of 7 days. After administration of the substance, all animals exhibited a definite decrease of mobility, which was reversible within 7 days. A brown coloration of the liver was found on dissection of the animals that died (Maickel and McFadden, 1979).

For the intensity of hypnotic action (loss of postural reflex) of 2-methylpropanol-1 on intraperitoneal injection (male Swiss-Webster mice, 5–8 weeks old) an ED_{50} of 14.9 mmol/kg was found (corresponding to approx. 1.1 g/kg; no further details of tests; Lyon et al., 1981).

The intravenous LD_{50} in female white mice (breed H, 7–8 weeks old) was found to be 8.22 mmol/kg (corresponding to 610 mg/kg; Chvapil et al., 1962).

Single exposure to 1818 ppm for 5 minutes led in the mouse to a 5% reduction of breathing rate (RD_{50}). It was recommended that a safety factor of 10–100 be applied to this RD_{50} value in establishing a TLV value for irritant substances (de Ceaurriz et al., 1981).

Three mice were exposed between 3 and 30 times to 2-methylpropanol-1 (0.12 ml of 2-methylpropanol-1 vapourized in 5 liters of air), in each case until narcosis occurred after each exposure. After 14, 99 and 136 hours of exposure, histopathological examination of the animals revealed benign fatty degeneration in the liver, kidneys and heart. Another 6 mice were exposed to similar concentrations (0.1–0.12 ml/5 liters of air) between 6 and 25 times; the exposure time was 9 hours in each case; narcosis did not occur. The animals showed an increase in weight. Again only benign reversible fatty degeneration of the liver and renal parenchyma was found after 52, 79, 135, 177 and 223 hours of exposure (Weese, 1927).

The oral LD_{50} for rats (60–100 g weight, 10 days follow-up period, strain not stated) was 2.65 g/kg (1.79–3.99 g/kg) for males and 3.10 g/kg (1.97–4.88 g/kg) for females. Fatally poisoned rats generally died within 18 hours. On dissection and histological examination, the animals exhibited hyperaemia of the liver and fatty infiltration, swelling and necrosis of the kidneys (Purchase, 1969).

Smyth et al., (1954) determined the oral LD_{50} of 2-methylpropanol-1 in groups of 5 male and 5 female Carworth-Wistar rats (90–120 g) as 2.46 g/kg (1.60–3.78 g/kg). The follow-up period was 14 days.

For determination of acute inhalation toxicity of 2-methylpropanol-1, 10 male and 10 female Spraque-Dawley rats (185 g ± 15 g) inhaled an average concentration of 6.5 mg 2-methylpropanol-1 per liter of air for 4 hours. Observation time was 14 days. All the animals survived, and there were no symptoms of intoxication. The LC_{50} was therefore >6.5 mg/l (approx. 2100 ppm; BASF, 1979a).

In the Inhalation Hazard Test (OECD) two different samples of 2-methylpropanol-1 (purity not stated) were used. In this investigation, Sprague-Dawley rats (in each case 6 male and 6 female animals) inhaled, at 20° C, an atmosphere enriched with 2-methylpropanol-1. The follow-up period was 14 days. In the first test all 12 rats survived exposure for 7 hours. The animals exhibited dyspnoea, loss of the blinking reflex and narcosis, but after 1 day the animals displayed nothing unusual. In the second test 6 male rats and 6 female rats in each case survived exposure for 3 hours; with 7 hours exposure, 1 of the 6 animals died (sex not stated); in these animals there was narcosis, eyelid closure and watery nasal secretion. Macroscopic examination of the animal that died revealed a grey coloration of the liver. The animals that survived did not exhibit any pathological changes macroscopically at the end of the observation period (BASF, 1978a, 1979b).

In the range-finding test, 6 rats inhaled a nominal concentration, not determined analytically, of 8000 ppm for 4 hours; 2 animals died. With a saturated vapour/air mixture, 2 hours was the maximum exposure period that could be survived (individual details not given; Smyth et al., 1954).

When undiluted 2-methylpropanol-1 was aspirated artificially by anaesthetized rats at a dose of 0.2 ml/animal, 10/10 animals immediately stopped breathing and died; if a solution (85%) of the alcohol was aspirated, 3/10 rats survived (Gerarde et al., 1966).

After intraperitoneal administration in rats the dose of 2-methylpropanol-1 leading to ataxia (graded 3 on a 7-point scale of increasing severity of neurological symptoms) was 54 mmol/kg (corresponding to approx. 4 g/kg; Shoemaker, 1981).

The oral LD_{50} of 2-methylpropanol-1 in the rabbit (both sexes, 1.5 to 2.5 kg) was 41 mmol/kg (corresponding to 3.04 g/kg). The follow-up period was 24 hours. The narcotic dose ND_{50}, defined as the dose that caused stupor and loss of voluntary movements in half of the rabbits, was also determined. The ND_{50} was 19 mmol/kg (corresponding to 1.41 g/kg). In this test, high doses (no figures given) caused loss of the corneal reflex, nystagmus, bradycardia and dyspnoea (Munch, 1972).

The test substance was administered by stomach tube to rabbits (1000–2500 g, no details given of breed, sex and number of animals per group). The minimum narcotic dose (MND; the dose causing mild narcosis in half the animals, as shown by stupor, immobility, lying on the side or stomach, short-term resumption of movement and standing up on manual compression or stimulation) was 1.75 ml/kg. The lethal dose was given as 3.75 ml/kg (Munch and Schwartze, 1925).

The dermal LD_{50} for the rabbit was 4.24 ml/kg (2.52–7.12 ml/kg). The test substance was applied occlusively to the shaved skin for 24 hours; the follow-up period was 14 days (Smyth et al., 1954).

After intravenous administration in the ear vein of 3 rabbits (2.5–3.7 kg), the minimum anaesthetizing dose of 2-methylpropanol-1 (causing loss of corneal reflex) was found to be 0.93 g/animal, and the minimum lethal dose was 2.64 g/animal (Lehmann and Newman, 1937).

The lethal dose of 2-methylpropanol-1 for the cat after intravenous administration was found to be 0.9 ml/kg (injection as 5%

solution in physiological NaCl solution, 2 ml/minute, investigation of 1 cat; Macht, 1920).

7.3 Skin and mucous membrane effects

In 2-hour patch tests on the shaved dorsal skin of white rabbits, undiluted 2-methylpropanol-1 caused slight reddening with subsequent peeling of the skin (BASF, 1978b).

Undiluted 2-methylpropanol-1, applied occlusively for 24 hours to the shaved skin of 5 albino rabbits, produced capillary irritation that was just visible (Smyth et al., 1954).

In 6 male rabbits, 0.5 ml of 2-methylpropanol-1 proved to be "moderately irritant" when applied occlusively for 24 hours to the shaved intact and abraded skin (BASF, 1978b; Fraunhofer, 1979a).

Instillation of 0.005 ml into the rabbit eye caused severe irritation with damage to the conjunctiva and cornea (Smyth et al., 1954).

In 6 male albino rabbits (white New Zealanders), instillation of 0.1 ml undiluted test substance into the conjunctival sac produced slight to moderate irritation of the conjunctiva (BASF, 1978; Fraunhofer, 1979b).

Carpenter and Smyth (1946) investigated the irritant effect of 2-me-thylpropanol-1 on the eye of white albino rabbits, and graded it 5 on a 10-point scale of increasing irritation.

7.4 Sensitization

No information available.

7.5 Subchronic/chronic toxicity

Male Wistar rats (4 months old, number not given) received a 1 M solution (74.12 g/l) of 2-methylpropanol-1 as drinking water for 4 months. The intake of 2-methylpropanol-1 was said to be 6.5 nmol/100 g/day (test duration 15 days) and 12.6 nmol/100 g/day (test duration 120 days). The amount of test substance ingested would thus be very small, presumably there is a misprint in the quantities stated in the publication. On the assumption that rats weighing from 250 to 400 g drink 20 ml of water daily, at a concentration of 74.12 g/l the daily intake would be about 3.75 to 6 g/kg. There were no changes in food intake and body-weight gain. On dissection, the stomachs were found to be enlarged with gas or full of food, and in some animals haemorrhages in the small intestine and constipation were observed. Changes in the liver (fatty degeneration, cirrhosis) were not seen (Hillbom et al., 1974).

7.6 Genotoxicity

With concentrations of 3–5% 2-methylpropanol-1, various strains of E. coli were killed to a large extent in approx. 35 minutes at 37° C. At a concentration of 2.5%, a survival rate of the E. coli strains of 1–10% was considered probable. With these pretreated E. coli CA 274 bacteria (0.1 ml/plate), an increased rate of reverse mutations was found in comparison with the spontaneous rates seen in corresponding tests with untreated bacteria (Hilscher et al., 1969).

7.7 Carcinogenicity

In a carcinogenicity study which does not meet present-day requirements, 2-methylpropanol-1 (degree of purity doubly distilled, no further information) was administered to male and female Wistar rats (colony-bred), twice per week, either orally (by stomach tube) or subcutaneously. It is not stated whether the alcohol was administered undiluted or in a preparation. 30 Wistar rats in each case (30 animals with oral administration, 30 animals with subcutaneous administration, no information on sex distribution) served as controls and received physiological NaCl solution. The animals were observed until they died naturally, then dissected and prepared histologically; the blood picture (leucocytes, differential blood count) was also analysed. Specific details of the results of these investigations were not given, or only in general terms. The following were listed as chronic-toxic lesions in the treated animals: Hyperplasia of the blood-forming parenchyma of the bone marrow, leukaemic infiltration in the liver and kidneys (all 3 haematopoietic cell systems were affected), toxic liver damage (congestion, cell necrosis, fibrosis, cirrhosis), localized scar formation in the heart muscle, occasionally pancreatitis and fibrosis of the pancreas. A summary of the tumours observed is given in the following table (page 51).

The absolute number of tumours was increased, compared to the controls, with both oral and subcutaneous administration; the tumours were distributed among different organ systems (Gibel et al., 1974, 1975).

Criticisms of the study referred to include:
- There is insufficient information on the purity of the alcohol used.
- The small number of animals limits interpretation of these results. No details are given of tumours in historical controls.
- There are no range-finding studies.
- Only one dose was administered, therefore it is not possible to establish a dose-response relationship.

Table 1. Test methods and results of the investigation into the carcinogenic effect of 2-methylpropanol-1 (Gibel et al., 1974, 1975)

Group	Route of admini-stration	Number of animals/group	Dose ml/kg	Mean total dose (ml)	Mean sur-vival time (days)	Animals with tumors total		malignant		benign	
						N	%	N	%	N	%
Control*	oral	25	1.0[5]	—	643	3	12	—	—	3[1]	12
Control*	s. c.	25	1.0[5]	—	643	2	8	—	—	2[1]	8
2-methyl-propanol-1	oral	19	0.2[5]	29.0	495	12	63	3[3]	16	9[2]	47
2-methyl-propanol-1	s. c.	24	0.05[5]	9.0	544	11	46	8[4]	33	3[2]	13

* 0.9% NaCl
[1] Papillomas or papillomatoses of the forestomach, fibroadenomas of the mammary gland
[2] Not defined for this group
[3] 1 animal with forestomach carcinoma and hepatocellular carcinoma, 1 animal with forestomach carcinoma and myeloid leukaemia, 1 animal with myeloid leukaemia
[4] 2 animals with forestomach carcinomas, 2 animals with liver sarcomas, 1 animal with spleen sarcoma, 1 animal with mesothelioma, 2 animals with retroperitoneal sarcomas
[5] Twice per week

- The dose used in each case produced chronic-toxic lesions, so it may be assumed that the dose chosen was above the usual doses for carcinogenicity tests.
- The documentation of the global findings does not permit evaluation.

Consequently, this study cannot be evaluated.

7.8 Reproductive toxicity
No information available.

7.9 Effects on the immune system
No information available.

7.10 Other effects
No information available.

8. Experience in humans

In a drinking test (no precise details of the method), test subjects (number not specified) drank 40% ethanol (by volume) and 3.75 mg/l of 2-methylpropanol-1 in orange juice (total quantity not stated). 2-methylpropanol-1 occurs in varying amounts in alcoholic drinks. Isobutyraldehyde and isobutyric acid were detected as metabolites in the blood. The highest blood concentrations for 2-methylpropanol-1 were found immediately after drinking (approx. 3 µmol/l), and 9 hours later the figure was 1 µmol/l. The isobutyraldehyde concentration rose continuously in the period of investigation (approx. 3 µmol/l after drinking, approx. 6 µmol/l 9 hours later). The concentration of isobutyric acid decreased from 18 µmol/l immediately after drinking to approx. 11 µmol/l 9 hours later. The maximum concentration of 2-methylpropanol-1 in the urine was reached 1 hour after administration of the alcohol (130 µmol/l), isobutyraldehyde rose from approx. 5 to 7 µmol/l from the 1st to the 9th hour, and the concentration of isobutyric acid decreased from approx. 90 µmol/l (at the end of oral administration) to approx. 25 µmol/l after 9 hours. In addition, propionaldehyde (increase from approx. 6 to 9 µmol/l within 9 hours), propionic acid (approx. 50 µmol/l after 1 hour – approx. 60 µmol/l after 8 hours – approx. 40 µmol/l after 9 hours) and succinic acid (30 µmol/l after 1 and 8 hours – 25 µmol/l after 9 hours) were detected. Elimination is said to be effected mainly by conjugating with glucuronic acid (no further details; Rüdell et al., 1983).

The same investigators conducted drinking tests on 10 test subjects, who were administered 40% ethanol (by volume) in orange juice alone and with the addition of isobutanol (1 g/l; the quantity drunk was not stated). The following tests were undertaken before the start of drinking and 1 and 9 hours after the end of drinking: testing of reaction times, two-handed coordination tests, d2-tests (attention loading tests), and subjective judgement. In the acute alcoholization phase there was good correlation between the expected deterioration in performance and the blood ethanol concentration. In the post-alcoholic phase, a considerable increase in the number of errors of the subjective symptoms of hangover were recorded in tests with higher alcohols, including 2-methylpropanol-1 (Rüdell et al., 1981).

According to Salo (1970), the odour threshold for 2-methylpropanol-1 in a test fluid (94.4% alcohol distilled from grain) was 75 ppm.

Amoore and Hautala (1983) gave 1.6 ppm as the odour threshold of 2-methylpropanol-1 in the air. Eye lesions have been observed in workers in a wire enamelling works, processing an enamel containing butyl acetate and 2-methylpropanol-1 as solvent and diluent. The workers complained of a burning sensation in the eyes, tickling and burning in the throat, and loss of appetite. After rather severe conjunctival irritation, 8 workers also suffered disturbances of vision, caused by the formation of vacuoles in the epithelium of the cornea. The objective and subjective symptoms disappeared after a few days. In two patients, there was a relapse after the work was resumed (Büsing, 1951).

References

Amoore, J.E., Hautala, E.
Odor as an aid to chemical safety: Odor thresholds compared with threshold limit values and volatilities for 214 industrial chemicals in air and water dilution.
J. Appl. Toxicol., 3, 272–290 (1983)

BASF AG
Gewerbehygiene und Toxikologie
Akutes Inhalationsrisiko – Inhalations-Risiko-Test (Ratte) (1978a)

BASF AG
Gewerbehygiene und Toxikologie
Akute Hautreizwirkung von iso-Butanol am Kaninchen (1978b)

BASF AG
Gewerbehygiene und Toxikologie
Bericht über die Bestimmung der akuten Inhalationstoxizität LC_{50} von i-Butanol bei 4stündiger Exposition an Sprague-Dawley-Ratten (1979a)

BASF AG
Gewerbehygiene und Toxikologie
Bericht über die Prüfung der akuten Inhalationsgefahr (akutes Inhalationsrisiko) von i-Butanol, Prod. Nr. 00902 an Sprague-Dawley-Ratten (1979b)

Büsing, K.-H.
Augenschädigungen durch Butylazetat und Isobutylalkohol in einem Kabelwerk
Zentralbl. Arbeitsmed. Arbeitsschutz, 2, 13–14 (1951)

Carpenter, C.P., Smyth, H.F.
Chemical burns of the rabbit cornea
Am. J. Ophthalmol., 29, 1363–1372 (1946)

de Ceaurriz, J.C., Micillino, J.C., Bounet, P., Guener, J.P.
Sensory irritation caused by various airborne chemicals
Toxicol. Lett., 9, 137–143 (1981)

Chvapil, M., Zahradnik, R., Cmuchalowa, B.
Influence of alcohols and potassium salts of xanthogenic acids on various biological objects
Arch. Int. Pharmacodyn. Ther., 135, 330–343 (1962)

Clegg, D.J.
The hen egg in toxicity and teratogenicity studies
Food Cosmet. Toxicol., 2, 717–727 (1964)

Deutsche Forschungsgemeinschaft
Maximum concentrations at the workplace and biological tolerance values for working materials (1986)
Report no. XXIII. Commission for the investigation of health hazards of chemical compounds in the work area.

Dillingham, E.O., Mast, R.W., Bass, G.E., Aufian, J.
Toxicity of methyl- and halogen-substituted alcohols in tissue culture relative to structure-activity models and acute toxicity in mice
J. Pharm. Sci., 62, 22–30 (1973)

Fraunhofer-Institut für Aerobiologie, Grafschaft
Bericht über die Prüfung von i-Butanol 78/306 auf primäre Hautreizwirkung
Investigation for BASF AG (1979a)

Fraunhofer-Institut für Aerobiologie, Grafschaft
Bericht über die Prüfung von i-Butanol 78/306 auf Schleimhautreizwirkung
Investigation for BASF AG (1979b)

Gaillard, D., Derache, R.
Métabolisation de différents alcools présents dans les boissons alcooliques chez le rat
Traveau de la société de pharmacie de Montpellier, 25, 51–62 (1965)

Gaillard, D., Derache, R.
Action de quelques alcools aliphatiques sur la mobilisation de différentes fractions lipidiques chez la rate
Food. Cosmet. Toxicol., 4, 515–520 (1966)

Gerarde, H.W., Ahlstrom, D.B., Linden, N.J.
The aspiration hazard and toxicity of a homologous series of alcohols
Arch. Environ. Health, 13, 457–461 (1966)

Gibel, W., Lohs, K.H., Wildner G.P.
Experimentelle Untersuchung zur kanzerogenen Wirkung höherer Alkohole am Beispiel von 3-Methylbutanol-1, Propanol-1 und 2-Methylpropanol-1
Z. Exp. Chir., 7, 235–239 (1974)

Gibel, W., Lohs, K.H., Wildner, G.P., Schramm, T.
Experimentelle Untersuchung zur kanzerogenen Wirkung von Lösungsmitteln am Beispiel von Propanol-1, 2-Methylpropanol-1 und 3-Methylbutanol-1
Arch. Geschwulstforsch., 45, 19–24 (1975)

Hedlund, S.-G., Kiessling, K.-H.
The physiological mechanism involved in hangover 1. The oxidation of some lower aliphatic fusel alcohols and aldehydes in rat liver and their effect in the mitochondrial oxidation of various substrates
Acta Pharmacol. Toxicol., 27, 381–396 (1969)

Hillbom, M.E., Franssila, K., Forander, O.A.
Effect of chronic ingestion of some lower aliphatic alcohols in rats
Res. Commun. Chem. Pathol. Pharmacol., 9, 177–180 (1974)

Hilscher, H., Geissler, E., Lohs, K.H., Gibel, W.
Untersuchungen zur Toxizität und Mutagenität einzelner Fuselöl-Komponenten an E. Coli
Acta Biol. Med. Germ., 23, 834–852 (1969)

Kühn-Birett
Merkblätter Gefährliche Arbeitsstoffe, U 38
Ecomed-Verlagsgesellschaft (1980)

Kühnholz, B., Wehner, H.-D., Bonte, W.
In-vitro Untersuchungen zur Löslichkeit aliphatischer Alkohole in Körpergeweben
Blutalkohol, 21, 308–318 (1984)

Lehmann, A.J., Newman, H.W.
Comparative intravenous toxicity of some monohydric saturated alcohols
J. Pharmacol. Exp. Ther., 61, 103–106 (1937)

Lyon, R.C., McComb, J.A., Schreurs, J., Goldstein, D.B.
A relationship between alcohol intoxication and the disordering of brain membranes by a series of short-chain alcohols
J. Pharmacol. Exp. Ther., 218, 669–675 (1981)

Macht, D.J.
A toxicological study of some alcohols with especial reference to isomers
J. Pharmacol. Exp. Ther., 16, 1–10 (1920)

Maickel, R.P., McFadden, D.P.
Acute toxicity of butyl nitrites and butyl alcohols
Res. Commun. Chem. Pathol. Pharmacol., 26, 75–83 (1979)

Munch, J.C.
Aliphatic alcohols and alkyl esters; narcotic and lethal potencies to tadpoles and to rabbits
Ind. Med., 41, 31–33 (1972)

Munch, J.C., Schwartze, E.W.
Narcotic and toxic potency of aliphatic alcohols upon rabbits
J. Lab. Clin. Med., 10, 985–996 (1925)

Purchase, I.H.F.
Studies in kaffircorn malting and brewing XXII. The acute toxicity of some fusel oils found in Bantu beer
S. Afr. Med. J., 54, 795–798 (1969)

Rüdell, E., Bonte, W., Sprung, R., Frauenrath, C., Küssner, H., Sellin, J.-H.
Pharmakologische Wirkungen geringer Dosen höherer aliphatischer Alkohole
Blutalkohol, 18, 315–325 (1981)

Rüdell, E., Bonte, W., Sprung, R., Kühnholz, B.
Zur Pharmakokinetik der höheren aliphatischen Alkohole
Beitr. Gerichtl. Med., 41, 211–218 (1983)

Salo, P.
Determining the odor thresholds for some compounds in alcoholic beverages
J. Food Sci., 35, 95–99 (1970)

Shoemaker, W.
The neurotoxicity of alcohols
Neurobehav. Toxicol. Teratol., 3, 431–436 (1981)

Smyth, H.F., Carpenter, C.P., Weil, C.S., Pozzani, U.C.
Range-finding toxicity data, List V
Arch. Ind. Hyg. Occup. Med. 10, 61–68 (1954)

Ullmanns encyclopaedia of technical chemistry.
4th Edition, Volume 9, p.25
Verlag Chemie, Weinheim (1975)

Weese, H.
Vergleichende Untersuchungen über die Wirksamkeit und Giftigkeit der Dämpfe niederer aliphatischer Alkohole
Arch. Exp. Pathol. Pharmakol., 135, 119–130 (1927)

Windholz, M. (ed.)
The Merck Index, 10th Edition, p.739.
Merck & Co. Inc. Rahway, USA (1983)

Chloroacetamide

1. Summary and assessment

No information is available relating to toxicokinetics and metabolism. The oral LD_{50} for the rat lies between 70 and 138 mg/kg.

According to the available data from animal experiments, no localized irritative effects on skin or mucous membranes and no sensitizing effects are seen after a single application of chloroacetamide; however, the dilutions used for assessing these effects may not provide an adequate test. On the other hand, inflammatory degenerative skin changes develop following repeated application to the skin of the rabbit.

Paralysis of the extremities as well as histologically confirmed damage to the liver and heart (fatty infiltration) and also to the pancreas (deposition of haemosiderin) are seen following single and repeated (30 times) administration either parenterally or cutaneously, using corresponding doses. The "no observed effect level" lies between 10 mg/kg (subcutaneous or intravenous) and 50 mg/kg (dermal). In the 90-day feeding trial with a feed concentration of 500 ppm (approx. 50 mg/kg daily), delayed weight increase, leucocytosis and, in male rats, testicular damage (spermatogenic cyst-formation) are observed, which after a 4-week period of follow-up observation are not fully reversible. Concentrations of 100 ppm (10 mg/kg daily) and 20 ppm (2 mg/kg daily) do not lead to any corresponding changes.

No indications of a teratogenic effect of chloroacetamide are available: tested only on single days (days 7, 11, 12, 13 and 14) of gestation.

A mutagenic effect of chloroacetamide cannot be inferred from the available test results (Ames test, chromosome aberration test in vivo, dominant lethal test).

No research findings on carcinogenic effects are available.

Chloroacetamide can produce contact eczema in humans.

2. Name of substance

2.1 Usual name   Chloroacetamide
2.2 IUPAC-name   2-Chloroacetamide
2.3 CAS-No.   79-07-2

Chloroacetamide

3. Synonyms

Common and trade names
α-Chloroacetamide
Chloroacetic acid amide
Microcide Mergal AF

4. Structural and molecular formulae

4.1 Structural formula

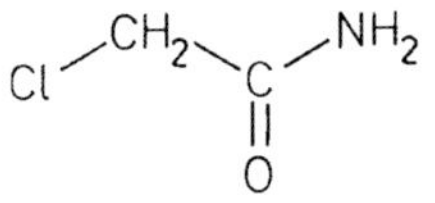

$$Cl-CH_2-\underset{\underset{O}{\|}}{C}-NH_2$$

4.2 Molecular formula C_2H_4ClNO

5. Physical and chemical properties

5.1 Molecular mass — 93.5 g/mol
5.2 Melting point — 121° C
    114–118° C
5.3 Boiling point — 224.5° C (at 993,51 hPa)
    Decomposition temperature above melting point
5.4 Density — no information available
5.5 Vapour pressure — 0.07 hPa (at 20° C)
5.6 Solubility in water — good solubility, 90 g/l (at 20° C)
5.7 Solubility in organic solvents — readily soluble in ether, dimethyl sulphoxide and dimethyl formamide, 1:10 in absolute alcohol, in isopropanol 40 g/l (at 20° C), in butanol 35 g/l (at 20° C)
5.8 Solubility in fat — no information available
5.9 pH-value — no information available
5.10 Conversion factor — 1 ppm $\hat{=}$ 3,8 mg/m³
    1 mg/m³ $\hat{=}$ 0.263 ppm
    (Weast, 1981–82; Hoechst; v. Kreybig et al., 1969; Hoechst, 1977, 1981).

6. Uses

As a preservative and as a catalyst in the manufacture of melamine and urea.

7. Results of experiments

7.1 Toxicokinetics and metabolism
No information available.

7.2 Acute and subacute toxicity
The following data on acute toxicity are given in the literature:

Table 1. Acute toxicity of chloroacetamide

Species	Route of administration	LD_{50} mg/kg	References
Rat	Oral	17.5	Hoechst, 1981
Rat	Oral	10.3	v. Kreybig et al., 1969
Rat	i. p.	19.2	Thiersch, 1971
Mouse	Oral	29.5	Hoechst, 1981
Mouse	Not stated	3.6	Chung, 1978
Mouse	i. p.	9.4	NIOSH, 1980
Mouse	i. v.	28.6	NIOSH, 1980
Rabbit	Not stated	30.4	Chung, 1978
Dog	Not stated	30.7	Chung, 1978

No information is given on symptoms of intoxication, dissection, or follow-up period. In some cases the cited references do not state the route of administration.

In a 30-day test, 5 rabbits ("Gelbsilber") received 2, 10, 25, 50 and 100 mg/kg (control 1 ml/kg 0.9% saline solution) subcutaneously on 30 successive days. Body weights were measured 3 times weekly. Before the start of the experiment and after the last administration, the blood profile (Hb, erythrocytes, leucocytes, differential blood count, Heinz bodies) and urine (protein, sugar, sediment) were investigated. Heart, lung, liver, spleen, kidneys, adrenal glands, ovaries, pancreas, brain, hypophysis and thyroid were examined histologically. Results are given in Table 2.

Table 2

Dose	Mortality	Findings
2 mg/kg	0/5	No findings
10 mg/kg	0/5	No findings
25 mg/kg	1/5	Paralysis of rear extremities, diarrhoea, injection site swollen, decrease of Hb
50 mg/kg	5/5[a]	Paralysis of front and rear extremities, swelling of skin of the neck, diarrhoea, no changes histologically
100 mg/kg	5/5[b]	No symptoms of intoxication
Control	0/5	No findings

[a] After 7–13 injections. [b] After 1–2 injections

Thus, when applied subcutaneously, chloroacetamide led to dose-dependent paralysis effects and, at 25 mg/kg or more, to death of the test animals. A damaging effect on the blood profile cannot be excluded (Hoechst, 1967a).

In a corresponding 30-day test (test conditions, parameters investigated, histology etc. as in subcutaneous application), chloroacetamide was administered intravenously to rabbtis on 30 successive days. The results are given in Table 3.

Table 3

Dose	Mortality	Findings
2 mg/kg	0/5	No findings
10 mg/kg	0/5	No findings
25 mg/kg	1/5	Paralysis of rear extremities starting from 13th injection, diarrhoea, weight loss, slight decrease of haemoglobin, organs histologically without findings
50 mg/kg	5/5[a]	Paralysis of extremities starting from 7th injection, diarrhoea, biochemical investigation not conducted, organs histologically without findings
100 mg/kg	5/5[b]	No symptoms of intoxication
Control[c]	0/5	No findings

[a] After 6–15 injections. [b] After 1–2 injections. [c] Physiological saline solution

With 30 intravenous injections in rabbits, chloroacetamide led to dose-dependent paralysis effects and, at 25 mg/kg or more, to death of the test animals (Hoechst, 1967b).

In another 30-day test with dermal application, chloroacetamide was applied to the shaved skin of the neck of rabbits on 30 successive days. To prevent licking, plastic collars were fitted round the animals' neck (test conditions, parameters investigated, histology etc. as in the 30-day tests with subcutaneous and intravenous injections). Chloroacetamide was applied in various doses as 40% suspension in water. The results see Table 4.

Table 4

Dose	Mortality	Findings
25 mg/kg	0/5	Incrustation and thickening of the application site
50 mg/kg	1/5	Incrustation and thickening of the application site
100 mg/kg	0/5	Incrustation and thickening of the application site, weight loss, fatty deposits in liver and heart, haemosiderin deposits in the spleen
200 mg/kg	0/5	Same changes as at 100 mg/kg
400 mg/kg	3/5[a]	Incrustation and thickening of the application site, weight loss, metabolic disturbances in liver and heart, haemosiderin deposits in the spleen
Control	0/5	No findings

[a] After 1–3 applications

Skin application of chloroacetamide 30 times in doses of 25 and 50 mg/kg (as a 40% aqueous suspension) was tolerated without systemic effects, apart from local changes of the skin. Starting at 100 mg/kg there was weight loss, with histological changes in the liver, heart and spleen. At 200 mg/kg, the same effects appeared, but with a more marked weight loss. 400 mg/kg proved lethal for 3/5 animals. Both, the local effects and the histological findings for the liver, kidneys and spleen were more pronounced than at the lower doses (Hoechst, 1967c).

7.3 Skin and mucous membrane effects

There are no investigations of skin or mucosal irritation by chloroacetamide. According to unpublished investigations, "CA 24" (a mixture of 70% chloroacetamide and 30% sodium benzoate) in 5% aqueous solution (3.5% chloroacetamide and 1.5% sodium benzoate) did not produce skin irritation in the Draize test (24-hour covered contact with the shaved and scarified skin of the rabbit; Sterner and Chibanguza, 1981a).

Moreover, 0.1 ml of a 5% aqueous solution of "CA 24" (composition as above) did not cause irritation of the mucous membranes of the rabbit eye (6 albino rabbits; Sterner and Chibanguza, 1981b).

7.4 Sensitization

In a sensitization test using the method of Magnusson and Kligman, chloroacetamide was tested on 20 guinea-pigs with corresponding control animals (Pirbright white). A 9% aqueous solution was used; in a preliminary test of primary irritant action this was tolerated without reaction, whereas 25 and 50% solutions and the undiluted product still caused erythema (24-hour contact). Testing for the development of skin sensitization was undertaken 14 days after the last treatment in the closed patch test with concentrations of 3, 1 and 0.3%. No signs of skin sensitization were seen (Sterner and Chibanguza, 1985).

In an open epicutaneous test, 10 female albino guinea-pigs (Pirbright white) were each treated locally on the shaved skin of the flank, 5 times weekly for 4 weeks, with a 3% or 1% chloroacetamide solution (no information on product purity or skin reactions). 10 days after the end of this induction phase, a 0.2% solution was applied to the skin of the opposite, untreated flank. Definite allergic reactions could not be produced in either group of previously-exposed guinea-pigs, when they were challenged with a 0.2% chloroacetamide solution (Schulz, 1984).

In addition, an investigation has been conducted considering the sensitizing effect of the preservative "CA 24" (70% chloroacetamide and 30% sodium benzoate) according to a modified Buehler method. 20 guinea-pigs were treated dermally with 0.5 ml of a 0.3% "CA 24" solution (concentration given by the commissioning party) once per week on the shaved skin of the left flank. The treated area of the skin was covered for 6 hours. 14 days after the last application, the 20 treated and 10 control guinea-pigs were treated with a 0.3% "CA 24" solution on the shaved skin of the right flank, and the skin reaction was compared with the controls after 2, 24 and 48 hours. No indications of skin sensitization were found. The report is only available in abstract form. Therefore fully assessment is not possible (Sterner and Chibanguza, 1981c).

7.5 Subchronic/chronic toxicity

In a 90-day feeding test (10 male and 10 female Wistar rats per test group, breed WISKF (SFP 71)), chloroacetamide was administered for 90 days in doses of 0, 20, 100 and 500 mg/kg feed (0, 2, 10 and 50 mg/kg body weight). Body-weight gain was only affected adversely in the group given the highest dose (500 mg/kg feed = 500 ppm) for male and female animals. At this dose, there was lower food intake in the first 30 days. Blood investigations, biochemical studies and analysis of the urine at the end of the test did not show any indications of chemically-induced changes in the highest test group, apart from reversible leucocytosis in male and female rats. In male rats at the highest dose, a significant decrease in the weight of the testes was observed at the end of the test, and this was still present after a 29-day follow-up period. There was a reduction in the size of the testes and epididymides. Immediately after the end of the test, histological examination showed depression or interruption of spermatogenesis and moderate proliferation of the interstitial Leydig cells. In the epididymides, neither mature spermatozoa nor precursors of them were seen, only cross-linked protein. At the end of the 29-day follow-up period there were signs of complete or incipient regeneration of the seminiferous tubules. No chemically-induced organ changes were seen in the groups given 100 and 20 mg/kg feed (about 10 and 2 mg/kg/day respectively; Hoechst, 1985).

In a 90-day study with dermal application, the commercially available preservative "CA 24" (mixture of 70% chloroacetamide and 30% sodium benzoate, no information regarding the purity of the two components) was applied dermally to the shaved skin of male rats (no details given regarding frequency of application per week). "CA 24" was applied to 10 rats in each case at 2% and 1% in Lanette and 2% in distilled water (corresponding to 1.4 and 0.7% chloroacetamide in Lanette and 1.4% in distilled water). The respective doses were 50 mg/kg (35 mg/kg chloroacetamide), 12.5 mg/kg (8.75 mg/kg chloroacetamide) in Lanette and 50 mg/kg (35 mg/kg chloroacetamide) in aqueous solution. No reasons were given for the doses chosen. A control group received Lanette. The parameters investigated, including body weight, food consumption, and histological examination of the testes and epididymides, did not reveal any pathological changes that could be attributed to the substance at the doses employed (Sterner et al., 1971).

7.6 Genotoxicity

In vitro. In the Salmonella/microsome test, chloroacetamide was not mutagenic to strains TA 98, TA 100, TA 1535 and TA 1537, at concentrations of 4–2500 µg/plate, with and without metabolic activation (S9-mix from Aroclor-induced rat liver). There is no information on whether the substance was investigated up to the toxic range (Hoechst, 1979).

In a Salmonella/microsome test (test strains not stated, report was only available as an abstract), the preservative "CA 24" (70% chloroacetamide, 30% sodium benzoate) was not mutagenic at concentrations of 0.5–1000 µg/plate (0.35–700 µg chloroacetamide/plate), either with or without S9-mix; toxic effects were seen in these bacteria at high doses (no precise details; Sterner, 1979a).

In vivo. In a report that is only available as a summary, it was found that two intraperitoneal injections of the preservative "CA 24" (70% chloroacetamide, 30% sodium benzoate) 24 hours apart, at doses of 50 mg/kg (roughly $1/3$ of LD_{50}, approx. 35 mg/kg chloroacetamide), 25 mg/kg (roughly $1/6$ of LD_{50}, approx. 17.5 mg/kg chloroacetamide) and 12.5 mg/kg (roughly $1/12$ of LD_{50}, approx. 8.75 mg/kg chloroacetamide), did not cause any structural or numerical chromosome aberrations in the bone marrow or the spermatogonia of Chinese hamsters; nor was an increase of micronuclei observed (there are no details of test procedure, evalutation etc.; Sterner, 1979b).

In a dominant lethal test on the mouse with the preservative "CA 24" (70% chloroacetamide, 30% sodium benzoate), 30 male mice each received a single intraperitoneal dose of 114 mg/kg (79.8 mg/kg chloroacetamide) or 123 mg/kg (86 mg/kg chloroacetamide). The controls received 10 ml/kg of the 0.5% methylcellulose used for making the suspension. The LD_{50} (24 hours) had been found to be 130 mg/kg (91 mg/kg chloroacetamide). The treated male mice were mated with 3 virginal female mice each week for 8 weeks. Implantations were reduced after the 1st, 2nd and 3rd weeks, and the fertility index was reduced after the 1st and 2nd weeks. The number of foetuses was also markedly lower after the 1st, 2nd and 3rd weeks, in comparison with the controls. In the opinion of the authors, these changes could be attributed to a toxic effect of "CA 24" in the first 3 weeks, which regressed after the 4th week. The number of resorptions, the mutation index and the number of dead foetuses remained unchanged throughout the test period, and from this the investigators conclude that a mutagenic effect should not be assumed (the report supplied does not contain detailed data; Sterner and Korn, 1979).

7.7 Carcinogenicity
No information available.

7.8 Reproductive toxicity

To test for its teratogenic potential, chloroacetamide was administered to rats (breed CD and BDIX, number not stated) in a single subcutaneous dose of 50 mg/kg, corresponding to about 71% of the LD_{50} (LD_{50} 70 mg/kg subcutaneous), on the 13th and 15th day of pregnancy. This dose had a toxic effect and led to the death of some of the young. There were no malformations. The surviving animals developed normally. No information was given on maternal toxicity (v. Kreybig et al., 1968, 1969).

When chloroacetamide (LD_{50} i.p. 50 mg/kg) was administered intraperitoneally to pregnant Long-Evans rats in a single dose of 20 mg/kg on the 7th day or the 11th and 12th days of pregnancy, no toxic or teratogenic effects on the foetuses were found in either group. The mothers similarly showed no symptoms of intoxication (no further details are given; Thiersch, 1971).

7.9 Effects on the immune system
No information available.

7.10 Other effects

After a single intraperitoneal injection of 75 mg/kg chloroacetamide to male Sprague-Dawley rats, there was a 90% decrease in the level of glutathione in the liver after 1 hour, but this was followed by a rapid increase to a level slightly above normal within 48 to 72 hours. 3 to 6 hours after administration of chloroacetamide there was development of lesions in the peripheral and intermediate region of the lobules of the liver (swelling and hydropic degeneration). At the same time, increased lipid peroxidation was seen (as measured by means of the thiobarbituric acid test). One week after treatment, the hydropic degeneration could still be detected in $^2/_3$ of all originally affected lobules of the liver. Fasting rats were more sensitive than non-fasting rats. A single intraperitoneal dose of 37.5 mg/kg chloroacetamide had an effect on the liver cell morphology in rats. At

112.5 mg/kg (intraperitoneal), there was a high mortality within 5 to 6 hours; histologically, the livers of these animals exhibited fatty degeneration, necrosis and leucocyte infiltration. Repeated intraperitoneal doses of 37.5 mg/kg chloroacetamide given on alternate days also led to liver swelling and hydropic degeneration within 2 weeks. These lesions were comparable to those of a single dose of 75 mg/kg (Anundi et al., 1980).

Chloroacetamide (0.2 mM) also induced GSH-depletion, lipid peroxidation and lysis of the liver cells in vitro. If the induced glutathione decrease was prevented e.g. with methionine, then lipid peroxidation and hepatocellular lysis did not occur either (Anundi et al., 1979).

8. Experience in humans

Chloroacetamide in the form of the preservative "CA 24" (70% chloroacetamide, 30% sodium benzoate), present in cosmetics and ointments, caused allergic contact eczema. One female patient had allergic reactions after using a body lotion (Nater, 1971).

Of 27 patients who were treated with a chloroacetamide-containing ointment, 17 developed allergic contact eczema (Smeenk and Prins, 1972).

Chloroacetamide was determined as the causative agent by means of a patch test. Sodium benzoate, which is also present in the preservative, proved to be non-sensitizing. Chloroacetamide is also contained in wallpaper adhesives, and accordingly cases of allergic contact eczema have also been seen in house painters. Two house painters who regularly worked with wallpaper adhesives over a period of 1 to 2 years developed contact dermatitis on the hands. The adhesives contained chloroacetamide and formaldehyde. In the patch test, both workers developed positive reactions to chloroacetamide (0.1%) and formaldehyde (2%; Bang-Pedersen and Fregert, 1976).

Of 2239 house painters in Stockholm, 190 developed various types of eczema. Of these, 5 (2.6%) gave positive reactions to chloroacetamide (Wahlberg et al., 1978; Högberg and Wahlberg, 1980).

Out of 47 patients with contact dermatitis on the hands, 5 (10.6%) reacted positively to this substance. These patients came into contact with adhesives containing chloroacetamide and formaldehyde, but their reaction to formaldehyde was negative (Wahlberg et al., 1978).

200 in-patients at a skin clinic, who suffered various skin diseases, mainly eczemas, were tested for skin-damaging properties by the epicutaneous patch test with "CA 24" (70% chloroacetamide, 30% sodium benzoate) as a 0.1% aqueous solution (0.07% Chloroacetamide); application sites were evalutated after 24, 48 and 72 hours. Skin reactions that would have indicated sensitization were not seen. 10 other test subjects with healthy skin were treated once daily on the same area of skin with a 0.1% "CA 24" solution for 14 days. No instances of irritant effects occurred (Röckl, 1970).

15 eczema patients, who were sensitized to tincture of benzoin, p-hydroxy-benzoates and balsam of Peru, were tested epicutaneously with a 0.1% "CA 24" solution (0.07% chloroacetamide and 0.03% sodium benzoate). At the dose chosen all tests were negative (Röckl, 1970).

In an clinical study of 296 patients (162 women, 134 men), who had been referred to an allergy clinic because of suspected contact allergy, chloroacetamide was tested as a 0.2% aqueous solution in the closed epicutaneous test (no information about the purity of the product is given). According to the investigator, the 0.2% chloroacetamide solution was well below the irritation threshold. 4 women and 3 men (previously treated with cosmetic and medical ointments) exhibited allergic reactions of the eczema type. The skin reactions found in these 7 patients were regarded as an expression of an earlier-acquired contact allergy to chloroacetamide (Schulz, 1984).

References

Amundi, I., Högberg, J., Stead, H.
Glutathione depletion in isolated hepatocytes: its relation to lipid peroxidation and cell damage
Acta Pharmacol. Toxicol., 45, 45–51 (1979)

Amundi, I., Rajs, J., Högberg, J.
Chloroacetamide hepatotoxicity: hydropic degeneration and lipid peroxidation
Toxicol. Appl. Pharmacol., 55, 273–280 (1980)

Bang-Pedersen, N., Fregert, S.
Occupational allergic contact dermatitis from chloroacetamide in glue
Contact Dermatitis, 2, 122–123 (1976)

Chung-hua I Hsueh Tsa Chih/Peking, 58, 462–466 (1978)
Studies on the snail extermination effect of chloroacetylamide and method for its detoxication
Cited in Chemical Abstracts, 92, 192639 J

Hoechst AG
Communication from Hoechst AG to the Employment Accident Insurance Fund of the Chemical Industry

Hoechst AG
Chloroacetamide, subcutaneous toxicity in rabbits, 30-day study
Unpublished investigation, Report 109/67 (1967a)

Hoechst AG
Chloroacetamide, intravenous toxicity in rabbits, 30-day study
Unpublished investigation, Report 110/67 (1967b)

Hoechst AG
Chloroacetamide, dermal toxicity in rabbits, 30-day study
Unpublished investigation, Report 111/67 (1967c)

Hoechst AG
Technical brochure Chloracetamide (1977)

Hoechst AG
Pharma Forschung Toxikologie
Ames-test with substance 128/79, Chloroacetamide
Unpublished investigation, report no. 351/79A (1979)

Hoechst AG
Material Safety Data Sheet Chloracetamide (1981)

Hoechst AG
Pharma Forschung Toxikologie
Chloroacetamide – 3 month feeding study (90 days) in the Wistar rat
Unpublished investigation, report no. 85.0899 (1985)

Högberg, M., Wahlberg, J.E.
Health screening for occupational dermatoses in house painters
Contact Dermatitis, 6, 100–106 (1980)

v. Kreybig, T., Preussmann, R., Schmidt, W.
Chemische Konstitution und teratogene Wirkung bei der Ratte
I. Carbonsäureamide, Carbonsäurehydrazide und Hydroxamsäuren
Arzneimittelforsch., 18, 645–657 (1968)

v. Kreybig, T., Preussmann, R., v. Kreybig, I.
Chemische Konstitution und teratogene Wirkung bei der Ratte
II. N-Alkylharnstoffe, N-Akylsulfonamide,N,N-Dialkylacetamide, N-Methylthioacetamid, Chloracetamid
Arzneimittelforsch., 19, 1073–1076 (1969)

Nater, J.P.
Allergic reactions due to chloroacetamide
Dermatologica, 142, 191–192 (1971)

Nater, J.P.
Chloroacetamide allergy
Dermatologica, 145, 403 (1972)

NIOSH (National Institute for Occupational Safety and Health)
Registry of Toxic Effects of Chemical Substances, 1980

Röckl, H.
Report on the investigation of the skin-damaging properties of the preservative CA 24
University clinic and polytechnic clinic for skin disease, Würzburg, 16.11.1970
Commissioned by Firma Biochema Schwaben, Memmingen

Schulz, K.H.
Allergological investigation of chloroacetamide
University skin clinic, Hamburg,18.12.1984
Commissioned by Firma Biochema Schwaben, Memmingen

Smeenk, G., Prins, F.J.
Allergic contact eczema due to chloroacetamide
Dermatologica, 144, 108–114 (1972)

Sterner, W.
Mutagenicity evaluation of preservative CA 24 in the Ames Salmonella/microsome plate test
IBR, Hannover, Project 0-0-513-78 (1979a)
Commissioned by Firma Biochema Schwaben, Memmingen

Sterner, W.
The action of "CA 24/Biochema Schwaben" on the chromosomes of the bone marrow and the spermatogonia of Chinese hamsters
IBR, Hannover, report dated October (1979b)
Commissioned by Firma Biochema Schwaben, Memmingen

Sterner, W., Chibanguza, G.
Skin irritation test in rabbits with the preservative CA 24
IBR, Hannover, Project 1-3-418-81 (1981a)
Commissioned by Firma Biochema Schwaben, Memmingen

Sterner, W., Chibanguza, G.
Eye irritation tests in rabbits with the preservative CA 24
IBR, Hannover, Project 1-3-417-81 (1981b)
Commissioned by Firma Biochema Schwaben, Memmingen

Sterner, W., Chibanguza, G.
Modified Buehler sensitization test with the preservative CA 24
IBR, Hannover, Project 2-5-419-81 (1981c)
Commissioned by Firma Biochema Schwaben, Memmingen

Sterner, W., Chibanguza, G.
2-Chloroacetamide — delayed contact hypersensitivity in Guinea-pigs, modified method of B. Magnusson and A.M. Kligman
Report IBR, Walsrode, Project No. 2-5-454-85 (1985)
Commissioned by Firma Biochema Schwaben, Memmingen

Sterner, W., Korn, W.-D.
Dominant lethal test in the mouse with the preservative CA 24
IBR, Hannover, Project 2-1-183-79 (1979)
Commissioned by Firma Biochema Schwaben, Memmingen

Sterner, W., Heisler, E., Messow, C., Wenzel, S.
3 month, dermal toxicity test of CA 24 in the rat
Report IBR, Anderten (1971)
Commissioned by Firma Biochema Schwaben, Memmingen

Thiersch, J.B.
Congenital Malformations of Mammals
P. 95–113
Masson und Cie, Paris 1971

Wahlberg, J.E., Högberg, M., Skare, L.
Chloroacetamide allergy in house painters
Contact Dermatitis, 4, 116–117 (1978)

Weast, R.C. (ed.)
CRC-Handbook of Chemistry and Physics
62nd edition, C-67
CRC-Press, Boca Raton, Florida 1981–82

4-Chloro-2-nitroaniline

1. Summary and assessment

The substance is rapidly distributed through the body following oral or intravenous administration, with no special affinity for particular tissues or organs. The whole-body half-life in the rat is one hour. 4-Chloro-2-nitroaniline (CNA) is broken down to (as yet unidentified) metabolites which are soluble in water or ether. Elimination is completed within 72 hours, mainly in the urine (approx. 75%); to a lesser extent in the faeces (approx. 10%). Elimination of radioactivity is almost complete after 3 days. There is no indication of accumulation in the body.

CNA shows no acute toxic effect in animal experiments (oral LD_{50} in the rat >5000 mg/kg).

No irritative effects on skin or mucosa have been observed.

According to the Salmonella microsome test, CNA is mutagenic following metabolic activation.

Further investigations on carcinogenic properties are currently being undertaken in the USA.

2. Name of substance

2.1	Usual name	4-Chloro-2-nitroaniline
2.2	IUPAC-name	1-Amino-4-chloro-2-nitrobenzene
2.3	CAS-No.	89-63-4

3. Synonyms

Common and trade names	2-Amino-5-chloro-nitrobenzene
	p-chloro-o-nitroaniline
	CNA

4. Structural and molecular formulae

4.1 Structural formula

[Structure of 4-chloro-2-nitroaniline: benzene ring with NH$_2$, NO$_2$, and Cl substituents]

4.2 Molecular formula

$C_6H_5ClN_2O_2$

5. Physical and chemical properties

5.1	Molecular mass	172.58 g/mol
5.2	Melting point	116–117° C
5.3	Boiling point	–
5.4	Density	1.37 g/cm^3 (at 20° C)
5.5	Vapour pressure	No information available
5.6	Solubility in water	Minimal (0.05 g/100 g)
5.7	Solubility in organic solvents	Ethanol, ether
5.8	Solubility in fat	No information available
5.9	pH-value	No information available
5.10	Conversion factors	1 ppm $\hat{=}$ 7.08 mg/m^3
		1 mg/m^3 $\hat{=}$ 0.141 ppm
		(Ullmann, 1979; Weast, 1982; Hoechst, 1986)

6. Uses

For the manufacture of dyes, intermediate product in the manufacture of pigments.

7. Results of experiments

7.1 Toxicokinetics and metabolism

Each of 3 male F-344 rats (weight 160–180 g) received ^{14}C-marked 4-chloro-2-nitroaniline (CNA) orally (by stomach tube) or intravenously. Administration: 1.9–2.3 µCi per dose and per animal. The rats received 0.788, 7.88 or 78.8 µmol CNA/kg (corresponding to 0.14, 1.36 or 13.55 mg/kg) by stomach tube; 7.78 μmol/kg was administered intravenously. No toxic effects were observed with these doses. After 15 and 45 minutes and 2, 7, 24 and 72 hours, radioactivity was determined in the blood, liver, musculature, fatty tissue, kidneys, skin, heart, lungs, brain, spleen, bladder, testes, stomach and contents, small intestine and contents, and colon and contents. In addition, radioactivity was determined in the urine and faeces after 7 and 72 hours. The elimination of radioactivity via the bile was determined on anaesthetized rats following intravenous administration of ^{14}C-marked CNA (7.88 μmol/kg corresponds to 1.36 mg/kg). Following oral administration, CNA was almost completely absorbed through the gastro-intestinal tract (>97%), the absorption rate having no correlation with the dose administered. The radioactivity distribution in the individual tissues was the same after 3 days, whether administered orally or intravenously. Following intravenous injection, CNA was distributed within 15 minutes in all the tissues examined. CNA therefore shows no organ-specific affinity whether administered orally or intravenously. CNA was rapidly metabolised, giving water-soluble or ether-extractable metabolites, the identity of which has not been determined. Elimination followed within 24–72 hours; with approx. 75% excreted in the urine and 10% in the faeces. Similar results were obtained after oral administration. Following intravenous administration of radioactively marked CNA into the bile, approx. 12% of the administered radioactivity was eliminated after 4 hours. The metabolites found in the faeces and the bile (HPLC analysis) were generally similar. The whole-body half-life of CNA was 1 hour; after 3 days the radioactivity in the body had almost completely disappeared. About 70% of the radioactivity eliminated through the urine was found as the sulphate-conjugate of a single metabolite, with the remaining 30% eliminated in the form of 7 additional metabolites and as trace amounts of the starting substance. No identification of the metabolites was carried out. There was no indication of any accumulation of CNA or of its metabolites and no indication at these dose levels of saturation of the pathways involved in the intake, distribution, metabolism or elimination of CNA (Chopade and Matthews, 1983).

7.2 Acute and subacute toxicity

After CNA was orally administered to female rats, it had an LD$_{50}$ of >5000 mg/kg (follow-up period: 14 days). The animals showed transient prostration (prone position) and their skin and urine took on a yellow color. The intravenous LD_{50} of CNA for mice is given as 63 mg/kg, although further information is lacking (Hoechst 1979b; Sax, 1979).

7.3 effects Skin and mucous membrane

No irritation of the skin was observed in rabbits after 24 hours covered application (assessed in accordance with the guidelines of the FDA; Hoechst, 1979a).

In the eye, CNA caused transient yellowing of the sclera and to part of the cornea, irritation of the conjunctiva was not seen (assessed in accordance with the guidelines of the FDA; Hoechst, 1979a).

7.4 Sensitization
No information available.

7.5 Subchronic/chronic toxicity
No information available.

7.6 Genotoxicity

To assess the mutagenic effect of CNA, the Salmonella/microsome test was used with and without metabolic activation (S 9-mix from the livers of male Sprague-Dawley rats and of Syrian hamsters induced with Aroclor 1254) with the strains Salmonella typhimurium TA 1535, TA 1537, TA 98 and TA 100. The test substance was said to be >99% pure. Doses of 3.3 to 1000 µg/plate (toxic dose) were used. CNA was mutagenic in strains TA 100 and TA 98 only with metabolic activation and the mutagenic effect was stronger with S 9-mix from the liver of hamsters (Haworth et al., 1983).

7.7 Carcinogenicity
No information available.

According to the National Toxicology Program 1987 in the USA, CNA is presently being tested for carcinogenicity in a long-term study.

7.8 Reproductive toxicity
No information available.

7.9 Efffect on the immune system
No information available.

7.10 Other effects
No information available.

8. Experience in humans

No information available.

References

Chopade, H.M., Matthews, H.B.
Disposition and metabolism of 4-chloro-2-nitroaniline in the male F 344 rat
J.Toxicol. Environ. Health, 12, 267–282 (1983)

Haworth, S., Lawlor, T., Mortelmans, K., Speck, W., Zeiger, E.
Salmonella mutagenicity test results for 250 chemicals
Environ. Mutagen. Suppl. 1, 3–142 (1983)

Hoechst AG
Haut- und Schleimhautverträglichkeit von p-Chlor-o-nitranilin, 4-chlor-2-nitro-anilin an Kaninchen
Unpublished report No. 212/79 (1979a)

Hoechst AG
Akute orale Toxizität von p-Chlor-o-nitranilin, 4-chlor-2-nitro-anilin an weiblichen Ratten
Unpublished report No. 213/79 (1979b)

Hoechst AG
Communication dated April 1986

National Toxicology Program
Public Health Service, Department of Health and Human Services, NTB-87-002 (1987)

NIOSH
Summary tables of biological tests
Nat. Research Council Chemical-Biological Coordination Center
6, 139 (1954) cited in Registry of Toxic Effects of Chemical Substances
NIOSH BX 1575000 (1975)

Sax, N.I. (ed.)
Dangerous properties of industrial materials
5th ed., p. 495
van Nostrand Reinhold Co., New York (1979)

Ullmanns Enzyklopädie der technischen Chemie
4th ed., vol. 17, p. 398
Verlag Chemie, Weinheim (1979)

Weast R.C. (ed.)
CRC Handbook of chemistry and physics
CRC Press, Boca Raton, 1981/1982

Ethylene thiourea

1. Summary and assessment _____

Ethylene thiourea (ETU) is quickly absorbed from the gastro-intestinal tract and can be found in the blood of rats within 5 minutes of oral administration; the half-life in the blood is 9.4 hours (rat) and 3.5 hours (cat). ETU is evenly distributed in the body tissues; there is, however, some accumulation in the thyroid gland. Excretion is principally in the urine; 70–80% of the administered substance or its metabolites are found in the urine of mice and rats, and 40–60% in the urine of monkeys within 48 hours. Apart from ETU itself, the chief metabolite found in the mouse is 2-imidazoline-2-yl-sulfenate, while in the rat imidazoline, ethyleneurea and imidazolone are found. In the cat, S-methyl-ETU, in particular, is found in addition to unchanged ETU and ethyleneurea. In the mouse, ETU is metabolized rapidly, and principally by the FMO (flavine-dependent monooxygenase) system and by cytochrome P 450.

ETU has a moderate acute toxicity (LD_{50} rat oral 545–1823 mg/kg). It can be absorbed through the skin. Exploratory tests on rats and guinea-pigs indicate that the local irritative action of this product on the skin is slight, but it has a sensitizing effect.

Repeated or prolonged administration of ETU produces enlargement of the thyroid with hyperplasia, reduction of coloid content and adenomatous growth (see below for carcinogenicity). The decrease in T3 and T4 and the increase in TSH in the serum as well as a reduced iodine-binding capacity indicate a dysfunction of the thyroid gland. Repeated administration leads to diminished food intake and body weight development as well as alopecia and salivation. Transient pathological changes in liver functions were observed, as were a rise in the liver weight and slight histological changes in the liver. A 28-day study of the effects on rats of inhaling an atmosphere containing dust of this substance is currently being carried out on behalf of the BG Chemie.

Subchronic and chronic exposure in rats points to a threshold dose of 25 ppm (90 days) and 5 ppm or <5 ppm (2 years).

The numerous findings from in vitro studies on the subject of the mutagenic effect of ETU on bacteria and yeast, or its harmful effects on DNA, have not been consistent. The test for sister chromatid exchange and for chromosome aberrations or use of the mouse lymphoma test in vitro, as well as all in vivo experiments to test for a genotoxic effect (drosophila, dominant lethal test, SCE, chromosome aberrations, micronucleus test) have provided no indication of a genotoxic effect. Only in the cell transformation test on renal cells of the baby hamster (BHK21) in the absence of any metabolic activation have positive effects been found in two independent studies. A definitive conclusion on the mutagenicity of ETU is, therefore, not yet possible. Currently, an UDS test on mouse liver cells and a HGPRT point mutation test on V 79 cells are being conducted on behalf of the BG Chemie.

ETU has a dose-related teratogenic and foetotoxic effect on rats and hamsters. The malformations effect chiefly the central nervous system and the skeleton. Malformations of the skeleton have been reported in mice too, although only after relatively high dosage. Malformations have also been observed in chick embryos following injection of ETU into the incubated egg. In rabbits, the substance produces an increase in the death rate of embryos, which show reduced brain weight and microscopically demonstrable kidney damage. Malformations, however, have not been observed. For guinea-pigs no teratogenic or foetotoxic effects after ETU administration have been observed according to the available findings. In cats the possibility of ETU having a teratogenic effect cannot be excluded. Malformations of the brain were also seen in the litters of four female animals exposed for only one day during the organ-forming period of gestation (rats, oral 10 mg/kg); no corresponding effect was seen with doses of 1, 3 and 5 mg/kg. According to present knowledge, ETU attacks the neuroblasts of the foetal brain (cell necrosis with resulting enlargement of the ventricles), whereby it is probably ETU itself and not one of its metabolites that exercises a direct teratogenic effect.

In carcinogenicity studies on rats, ETU produced dose-related thyroid carcinoma; this was already apparent after 6 months following high dosage and was accompanied or preceded by hyperplasia of the thyroid gland. The frequency of hepatoma was increased in mice on long-term application, whereas no tumours were observed in hamsters followed prolonged administration.

In humans, one case of a sensitizing effect of ETU following exposure has been described. Investigation of the T4 concentration in the serum of 5 workers who had been employed as mixers in the processing of ETU have shown that these values are lower than those in the control subjects. Disturbance of thyroid gland function following high exposure is therefore within the realms of possibility. A further, retrospective study of workers in the rubber industry has produced no grounds for suspecting the development of thyroid gland tumours in this field of industry. An inadequately conducted and documented retrospective study of women employed in the rubber manufacturing industry provides no indication of an increased frequency of congenital deformities among the women's children.

2. Name of substance

2.1 Usual name   Ethylene thiourea
2.2 IUPAC-name   2-Imidazolidinethione
2.3 CAS-No.   96-45-7

3. Synonyms

Common and trade names   ETU
                           4,5-Dihydro-2-mercaptoimidazole
                           Ethylenethiourea
                           N,N'-Ethylenethiourea
                           1,3-Ethylenethiourea
                           Imidazolidinethione
                           Imidazoline-2-thiol
                           Imidazoline-2(3H)-thione
                           Mercaptoimidazoline
                           2-Mercapto-2-imidazoline
                           Mercazin 1 NA 22
                           Pennac CRA
                           Rhenogram ETU
                           Rhodanin S 62
                           Soxinol 22
                           Thiourea, N,N'-(1,2-ethandiyl)-
                           Vulcacit NPV/C
                           Warecure C

Ethylene thiourea

4. Structural and molecular formulae

4.1 Structural formula

$$\underset{H}{\overset{H}{\underset{|}{N}}}\!\!\!\diagdown\!\!\!\underset{S}{\overset{}{\rightleftharpoons}}\;\;\underset{H}{\overset{}{N}}\!\!\!-\!\!SH$$

4.2 Molecular formula $C_3H_6N_2S$

5. Physical and chemical properties

5.1	Molecular mass	102.2 g/mol
5.2	Melting point	203–204° C
5.3	Boiling point	–
5.4	Density	1.4 g/cm^3 (at 20° C)
5.5	Vapour pressure	0.0027 hPa (at 100° C)
		0.029 hPa (at 130° C)
		0.088 hPa (at 151° C)
5.6	Solubility in water	2 g/100 ml (at 30° C)
5.7	Solubility in organic solvents	moderately soluble in methanol, ethanol, ethylglycol, pyridine; insoluble in acetone, ether, chloroform and benzene
5.8	Solubility in fat	no information available
5.9	pH-value	no information available
5.10	Conversion factor	1 mg/m^3 $\hat{=}$ 0.24 ppm
		1 ppm $\hat{=}$ 4.17 mg/m^3

(IARC, 1982; Römpp, 1977; Du Pont, 1973; Lewerenz and Plass, 1984)

6. Uses

Vulcanization accelerator and antioxidant in the rubber industry (IARC, 1982; Hoechst).

7. Results of experiments

7.1 Toxicokinetics and metabolism

^{14}C-ETU was administered orally to 2 female rhesus monkeys (40 mg/kg, 12.5 µCi/mg) and 4 female Sprague-Dawley rats (40 mg/kg, 0.75 µCi/mg). The animals were kept in metabolic cages for 48 hours. After 48 hours, the radioactivity was measured in the urine, in the faeces and in all tissues. The values found were given as a percentage of the administered radioactivity. In the monkeys, 48 hours after administration, 47% and 64% of the administered radioactivity was found in the urine, and 0.45% and 0.63% in the faeces. The radioactivity measured after 48 hours ("total tissues") was 28% and 21%, most being present in the muscles (7.8 and 15.4%), skin (1.6 and 3.1%) and blood (1.8 and 3.9%). Tissue concentrations were 0.14% ^{14}C/g in the thyroid, 0.024% ^{14}C/g and 0.014% ^{14}C/g in urinary bladder and gall bladder respectively, and between 0.004 and 0.007% ^{14}C/g in all other tissues. Accordingly, it did not appear that there was accumulation of ^{14}C-ETU in the thyroid of monkeys. In the rats (mean values from 4 rats), 81.9% ± 16% of the administered radioactivity was found in the urine, 1.32% ± 0.34% in the faeces and 0.73% ± 0.26% in the total tissue. In the thyroid of 2 animals there was 0.91% ^{14}C/g and 1.07% g/tissue; in the other 2 animals the concentration was 0.004% ^{14}C/g of tissue. In the liver there was 0.013% ^{14}C/g, in the other tissues 0.004% ^{14}C/g on average. According to these investigations, ETU was absorbed from the gastro-intestinal tract in both animal species. In the rats a high percentage of ^{14}C was excreted in the urine, and only a small residue remained in the tissues. On the other hand, in the monkeys about 25% of the radioactivity remained in the tissue under these test conditions (Allen et al., 1978).

Excretion of ETU was investigated by gas chromatography in 6 male Wistar rats and 6 male Hartley guinea-pigs (each 20 mg/kg p.o.). The animals were killed 96 hours after administration. Rats excreted approx 60% and guinea-pigs approx 44% of the dose in the urine within 24 hours. Rats eliminated 1.06% and guinea-pigs 0.78% in the faeces within 48 hours. In liver, kidneys, heart and muscle there was between 0.01 ppm and 0.086 ppm ETU after 96 hours. In the thyroid, in contrast, 0.83 ppm was found in rats, and 0.75 ppm in guinea-pigs (Newsome, 1974).

In another investigation, 3 male Sprague-Dawley rats received 4 mg/kg ^{14}C-ETU (26.4 µCi/mg) orally and 2 female cats received

4 mg/kg ^{14}C-ETU (2.4 µCi/mg) i.v.. Then the animals were kept in metabolic cages for 24 hours. Blood samples were taken hourly from the cats, from 15 minutes to 10 hours after administration. In the rats, 24 hours after administration, 82.7% of the administered radioactivity was found in the urine, and in the cat the figures were 74.7% and 86.6%. The half-life in the blood of the cats was found to be 3.5 hours. The following substances were determined chromatographically as metabolites in the urine: ETU, unchanged (rat 62.6%, cat 28%), S-methyl-ETU (rat 0%, cat 63.3%), ethylene-urea (rat 18.3%, cat 3.5%). In addition, imidazolone (4.9%) and imidazoline (1.9%) were found in the rat, but not in the cat. According to these investigations, within 24 hours the rat had excreted 82.7% of the radioactivity in the urine, and the cat 80.6% (Iverson et al., 1980).

In vitro, ^{14}C-ETU was metabolized by cat-liver and rat-liver microsomes to ethylene-urea, imidazoline and other unidentified compounds. S-Methyl-ETU was formed when ^{14}C-ETU was incubated with a cat-liver supernatant fluid containing S-adenosyl-methionine. It was deduced from these results that the cat can metabolize ETU extensively to the S-methyl derivative, and this might possibly explain the absence of a teratogenic effect in this species (Iverson et al., 1980).

On oral administration of a dose of 0.25 mM/kg of ^{14}C-ETU, 52% of the radioactivity was present in the urine as unchanged ETU after 24 hours, 12% as ethylene-urea and 36.7% as unidentified polar compounds (Jordan and Neal, 1979).

The distritution of ^{14}C-ETU (2-^{14}C-ETU and 4,5-^{14}C-ETU) in pregnant rats was investigated by Kato et al. (1976). Wistar rats (number not stated) received 100 mg ^{14}C-ETU/kg orally on the 12th day of pregnancy (45.87 µCi of 2-^{14}C-ETU or 30.45 µCi of 4,5-^{14}C-ETU). The animals were kept in metabolic cages. Up to 192 hours after administration, measurements were made of radioactivity in the blood and in the organs of the mothers and foetuses, first at hourly intervals and subsequently at longer intervals of time. ETU was absorbed rapidly after oral administration. The substance could be detected in the blood after just 5 minutes, the maximum being reached after 2 hours (0.48 µmol as 2-^{14}C-ETU/g), then the concentration decreased until 48 hours after administration, at which point radioactivity could no longer be detected in the blood. In 2 days, 82.5% of the administered radioactivity (4,5-^{14}C-ETU) was excreted in the urine, and 0.53% in the faeces. A small proportion of the radioactivity could be detected in the exhaled Co_2.

4,5-^{14}C-ETU was exhaled as radioactive CO_2 (up to 8 nmol as ^{14}C-ETU/2 minutes), whereas after administration of 2-^{14}C-ETU only traces (approx. 0.6 nmol as ^{14}C-ETU/2 minutes) of radioactive CO_2 were found. 4,5-^{14}C-ETU was bound to serum and to foetal cell constituents, but this was not seen with 2-^{14}C-ETU. It was deduced from these results that metabolism to CO_2 goes via fragmentation of the imidazolidine ring and decarboxylation of the 4 and/or 5C atom. In the organs of the mothers there was almost uniform distribution of radioactivity, which reached its maximum about 2 hours after administration (approx. 0.2 µmol as 2-^{14}C-ETU/g tissue). In contrast, there was accumulation of ETU in the thyroid (2 hours p.a.: 0.346 µmol as 2-^{14}C-ETU/g; 24 hours p.a.: 0.651 µmol as 2-^{14}C-ETU/g). In the foetuses, a concentration of 0.22 µmol as 2-^{14}C-ETU/g was measured 2 hours p.a., and after 24 hours there was 0.012 µmol as 2-^{14}C-ETU/g. ETU was accordingly distributed uniformly in the mothers and foetuses (Kato et al., 1976).

Comparable results were obtained by other researchers on oral administration to pregnant Wistar rats (50 µCi/kg ^{35}S-ETU) and Swiss white mice (100 µCi/kg ^{14}C-ETU) on the 15th day of pregnancy together with 240 mg ETU/kg. Three animals in each case were killed 3, 6, 12, 24 and 48 hours after administration, and the radioactivity in the liver, kidneys, muscles, blood, placenta, foetuses and urine was measured. In both species the ETU concentrations after 3 hours were roughly the same (150–180 µg/g tissue) in the tissues of the mothers and foetuses. After that (especially after 6 and 12 hours), there was a faster decrease of concentration in the mouse (mothers and foetuses; after 12 hours, approx. 10 to 13 µg ETU radioactivity/g tissue) than in the rat (approx. 60 to 80 µg/g tissue). After 24 hours the activities in the mouse, were between 12 and 17 µg/g tissue. In rats, 2–7 µg/g tissue of activity could still be detected in all tissues after 48 hours; in mice at the same time, there was still 2.9 µg/g tissue in the liver, but no further activity was measured in other organs. In the blood of the mothers the half-life for ETU and its metabolites was 5.5 ± 0.4 hours (mouse) and 9.4 ± 0.6 hours (rat). Thus, ETU was excreted more quickly by the mouse than the rat. Within 48 hours, 70.5% and 73.5% of the administered radioactivity had been excreted in the urine of the rat and mouse respectively, and the mouse excreted the activity more quickly than the rat. According to the chromatographic and radiochromatographic investigations that were conducted on the urine, the mouse evidently seemed to metabolize ETU by a different and faster route than the rat (Ruddik et al., 1976a, 1977).

Similar results were observed in another publication from the same team. ^{14}C-ETU was given orally to pregnant rats on the 11th, 12th or 15th day of pregnancy (240 mg/kg and 50 µCi/kg). The radioactivity was distributed uniformly between plasma and erythrocytes to the mother, and radioactive binding on the erythrocytes was reversible. The radioactivity was also uniformly distributed in the embryo, without indications of binding of ETU to DNA, RNA or protein (Ruddik et al., 1976b).

The principal metabolite of ETU (67 mg/kg orally as 2-^{14}C-ETU and as ^{35}S-ETU) in male NMRI mice found in the urine after appropriate preparation, in addition to unchanged ETU and other metabolites, was 2-imidazolin-2-yl-sulphenate; this was also detected in the serum and in the liver, but not in the faeces (Savolainen and Pyysalo, 1979).

Incubation of ETU solution with liver micro-somes from the mouse led to products that were identical to the metabolite in the living mouse. It was deduced from this that metabo-lism is localized in the microsomes (Savolainen and Pyysalo, 1979).

In another investigation of the metabolism of ETU, female Wistar rats (4 rats) received 200 mg/kg orally. The blood of the 4 animals was collected 2 hours after administration; the plasma was analyzed by chromatography (HPLC) and was found to contain 138 ppm ETU and, as a metabolite of ETU, 0.014 ppm 1-methylthiourea (Kobayashi et al., 1982).

In another investigation it was shown that the mouse (male and female Swiss-Webster) oxidized ETU preferentially via flavine-dependent monooxygenase (FMO), a microsomal NADPH-dependent enzyme. The very heat-sensitive FMO activity was significantly in the liver in 30-week-old male mice than in 2-week-old males. There was no difference in females. The in vitro-FMO increased metabolism of ETU was also lower in old male mice than in young males, but there was no such difference in old and young females. Similarly, binding of ^{14}C-ETU to protein of liver microsomes in vitro was significantly lower for old male mice than for young ones, whereas there was no such difference in female mice. In their discussion, the authors related the rapid FMO-dependent metabolism in the mouse to the low acute toxicity and slight or absent teratogenic effect of ETU in comparison with the rat; according to the authors, FMO-dependent hepatic binding of ETU might also play a part in the hepatocellular carcinomas observed in mice after chronic administration of ETU (Hui et al., 1988).

7.2 Acute and subacute toxicity

The data in the following table 1 are available concerning acute toxicity studies (follow-up periods are not stated). According to these studies, ETU is only of slight acute toxicity in animal tests; absorption through the skin is possible; salivation and weight loss are described as symptoms of intoxication (Khera, 1973).

In preliminary tests to investigate the embryotoxic effect of ETU, the following studies after repeated application were undertaken (Table 2; Chernoff et al., 1979).

ETU was administered orally by stomach tube for all animal species. There was no determination of food intake, weight gain, haematological or clinical-chemical parameters or histopathological changes.

7.3 Skin and mucous membrane effects

Four rats were each injected subcutaneously with 0.5 ml of a 20 or 10% solution of ETU in distilled water. The 20% solution caused marked hyperaemia at the injection site within 24 hours, whereas the 10% solution only caused slight hyperaemia within 24 hours (Matsushita et al., 1976).

The threshold value for a local irritant effect of ETU on the skin was determined on 6 female Hartley guinea-pigs weighing approx. 500 g. ETU was dissolved in distilled water and applied on the shaved skin of the flank under a bandage (for 24 hours). The threshold concentration for local irritation on the skin was found to be more than 10% ETU. No details of the test procedure are given (Matsushita et al., 1976).

No studies of the irritant effect of ETU on the eye are available.

7.4 Sensitization

Tests for possible sensitizing effect of ETU were conducted on 10 female Hartley guinea-pigs (300–500 g). The test method used was the maximization test of Magnusson-Kligman (with Freund's adjuvant). ETU was dissolved in distilled water. For induction of an allergic response, a 5% solution was injected intradermally and a 25% solution was applied locally. To provoke reaction, 0.5% and 2% solution was used locally. The allergic reaction was classified using a five – stage scheme – slight reaction (stage I) to extreme reaction (stage V). Provocation with 0.5% solution caused a stage II reaction in 1 out of 10 animals after 24 hours. After 48 hours, no animal showed a reaction. With provocation with the 2% solution, a stage IV reaction was found in 7 of the 10 animals after 24 hours, but after

Table 1. Acute toxicity of Ethylene thiourea

Species	Sex	Breed	Route of administration	LD$_{50}$ mg/kg	References
Rat	Male, Female	Wistar	Oral	940	Lewerenz and Plass, 1984
Rat	Male	Wistar	Oral	900 (680–1170)	Lewerenz et al., 1980
Rat	Male	Osborne Mendel	Oral	1832 (1379–2562)	Graham and Hansen, 1972
Rat	Female	Wistar	Oral	At 656 mg/kg death 25% (number not stated) 216 mg/kg tolerated	Khera, 1973
Rat	Female	Wistar	Oral	545	Teramoto et al., 1978
Mouse	Female	SLG-ICR	Oral	ca. 3000	Teramoto et al., 1978
Golden Hamster	Female		Oral	>3000	Teramoto et al., 1978
Rat	Female (pregnant)	Charles River	Cutaneous	>2250	Stula and Krauss, 1977

Mouse	Male		Oral	>1000	Doull et al., 1962
Mouse	Female	CD 1	Oral	5085	Plasterer et al., 1985
Mouse	Male, Female	Rehb: RIEMS/A	Oral	4000	Lewerenz and Plass, 1984
Rat	Female	Sprague-Dawley	80;40;20	15 days	25% lethality at 80 mg/kg
Mouse	Female	Charles River CD	300;200;100	10 days	No symptoms of intoxication
Hamster	Female		150;100;50	6 days	No symptoms of intoxication
Guinea pig		Hartley	100;50	19 days	No symptoms of intoxication

48 hours none of the guinea-pigs showed a reaction. In the opinion of the authors, ETU has a slight to moderate sensitizing effect (Matsushita et al., 1976).

7.5 Subchronic/chronic toxicity

Sprague-Dawley rats were administered ETU in the feed in concentration of 0, 1, 5, 25, 125 or 625 ppm (approx. 0.08, 0.40, 2.0, 10.0, 50 mg/kg bw/day). Groups of 20 males and 20 females received these concentrations in the feed for 30 days, 60 days or 90 days. 24 males and 24 females served as controls in each case. The following symptoms of intoxication were observed only in the 625 ppm groups starting from the 8th day of testing: salivation, hair loss, rough bristly coat, permanent hair loss and squamous skin. 14/40 rats died in the 625 ppm group (test duration 60 days, sex not stated). At 625 ppm (test duration 90 days) only 1 rat died. No deaths occurred at 125 ppm and below. Weight gain was only markedly reduced at 625 ppm in the feed, and was not adversely affected at 125, 25, 5, 1 and 0 ppm. Uptake of 125-iodine in the thyroid only showed a significant decline in the 625 ppm group on testing after 30, 60 and 90 days (30 days: control male 3.6%, female 3,5%, 625 ppm 1.2* and 2.1%* respectively; 60 days: control male 4.3%, female 3.5%, 625 ppm 1.9* and 2.4%* respectively; 90 days: control male 3.8%, female 4.1%, 625 ppm 2.5* and 3.7%* (*$P<0.05$)). In other dose groups, no impairment was seen compared with the controls. T3 and T4 were significantly reduced after intake of 625 ppm and sometimes even after 125 ppm for 30, 60 and 90 days, whereas TSH (only tested in the 30-day group) was significantly raised after feeding 625 and 125 ppm. The absolute weight of the thyroid and the relative weights of thyroid, brain, kidneys, testes and pituitary were increased in all animals of the 625 and 125 ppm groups after 30, 60 and 90 days. No changes occurred at lower concentrations. Macroscopically, enlarged red-coloured thyroids were seen in the 625 and 125 ppm groups, and alopecia in the high dose group. Without definite dose-dependence, centrilobular hyperaemiae were found in the liver down to 1 ppm. On histopathological examination of the thyroid, varying degrees of microfollicular hyperplasia were found above the 25 ppm group. After 30 days at 125 and 625 ppm, there were slight (slight to marked hypertrophy and hyperplasia) to moderate (marked hyperplasia and hypertrophy with reduced colloid content) changes in the thyroid, dependent on the dose. After 60 days the thyroids exhibited moderate changes (marked hyperplasia and hypertrophy with reduced colloid) in the 25, 125 and 625 ppm groups. After 90 days, the changes in the 625 ppm group ranged from marked microfollicular hyperplasia to solid adenomas; such changes were not seen at the lower doses. The livers were found to have slight centrilobular changes (swollen cytoplasm, indistinctly defined cell walls, cell strands disordered in the region of the central veins). For rats, the "no effect level" of ETU in the feed is regarded by the authors as 25 ppm over 90 days (Freudenthal et al., 1977).

Arnold et al. (1983) tested the reversibility of the thyroid changes induced by ETU. Groups of 50 male and 50 female Sprague-Dawley rats received 0, 75, 100 or 150 ppm ETU in the feed for 7 weeks. After this time 10 animals from each group were killed and examined. Further groups of 10, 10 and 20 animals were kept for a further 2, 3 and 4 weeks with ETU-free food and then investigated. For both sexes, food intake and mean body weight were reduced at all doses. There was a dose-dependent increase of mean thyroid weight (absolute and relative). The mean T4 blood values in the 150 ppm group (the only group in which they were measured) were significantly below the controls. In the follow-up periods with ETU-free feed, the measured parameters showed a tendency to return to normal. In another test, 5 groups of 20 male Osborne-Mendel rats received ETU in the feed for 30, 60, 90 or 120 days at concentrations of 0, 50, 100, 500 or 750 ppm. At the end of each test period, the animals were given 5 µCi of ^{131}-iodine (0.2 ml) i.p.. Half of the animals were killed 4 hours after injection, the remainder after 24 hours; the thyroids were removed and the radioactivity was determined. A significant decrease of body weight was seen after all feeding periods in the animals of the 500 and 750 ppm groups, and for the 100 ppm groups in the 60-day test. Food intake was significantly lower at concentrations in the feed of 100, 500 and 750 ppm after 30 and 90 days; after 60 and 120 days, only for animals of the 500 and 750 ppm group. In comparison with the controls, the relative weight of the thyroid was significantly higher after 30 and 60 days for the animals of 100, 500 and 750 ppm groups. After 90 days in the 500 and 750 ppm groups and after 120 days in the 50, 100, 500 and 750 ppm groups. Iodine uptake (cpm/mg tissue) was significantly lower in the 500 and 750 ppm groups 4 hours after injecting ^{131}-iodine in the thyroid after 30, 60, 90 and 120 days of ETU-diet. Measurement of iodine uptake 24 hours after injection of ^{131}-iodine additionally showed a significant decrease in the 100 ppm group with feeding for 30, 60 and 120 days. In the 50 ppm group, no significant difference from the controls was found at any point in time for any of the parameters investigated. Histologically, at 50 ppm there were no unusual findings in the thyroids, at 100 ppm there was hyperplasia, and at 500 and 750 ppm in addition to marked hyperplasia there was colloid deficiency and increased epithelium (Graham and Hansen, 1972).

In further tests, 68 male and 68 female Charles-River CD rats received 0, 5, 25, 125, 250 and 500 ppm ETU in the food for 2, 6 and 12 months (Graham et al., 1973) or 18 and 24 months (Graham et al., 1975). The rats were about five weeks old at the start of the tests. At the end of the 2-, 6- and 12-months feeding periods, 10 males and 10 females from each group were injected i.p. 0.2 ml physiological saline solution with approx. 5 µCi of ^{131}-iodine. At the end of the 18- and 24-months feeding period, 5 males and 5 females from each group were submitted to the same treatment. In each case the animals were killed after 24 hours of fasting, and the thyroid, heart, liver, kidneys, spleen and testes were removed. The organs were weighed and the content of ^{131}-iodine in the thyroid was determined. A haematological investi- gation was conducted on 10 rats from each group after 3 and 11 months (Graham et al., 1973) and after 17 and 22 months (Graham et al., 1975). With the exception of the 2-months test, the surviving animals were killed at the end of each test period, their organs were examined macroscopically and the thyroids were examined under the microscope. In addition, organs exhibiting severe lesions macroscopically were pre-pared for microscopic investigation (Graham et al., 1975). The body-weight gain of the male and female rats fed with ETU was significantly lower even at 25 ppm, in comparison with the control animals. Uptake of ^{131}-iodine into the thyroids was unchanged in all test groups after 6 months. After 12 and 18 months, in comparison with the controls, there was increased absorption of ^{131}-iodine for both sexes at 25 and 125 ppm in the feed; at 500 ppm there was decreased iodine absorption. There was no difference between male and female rats. After 24 months, iodine absorption in male rats was increased at 5 ppm, but it was decreased in the 500 ppm group. Only a tendency to decrease was observed for the female rats. The authors did not give a detailed assessment of the various increases or decreases of ^{131}-iodine absorption into the thyroid with administration of ETU; attention was mainly drawn to the reduced iodine binding capacity in the 500 ppm groups at various points in time. Haematological investigations (haemoglobin, haematocrit, leucocyte number, differential blood count) showed no differences between the control and test animals after 3 to 22 months. The relative weight of the liver was significantly higher in males after 2 and 6 months at 125–500 ppm, after 12 months at 125 ppm and after 18 months at 250 ppm. After 24 months there was no difference between groups. In the females, increased relative liver weights were found after 2 months at 125–500 ppm and after 6 and 12 months at 250 and 500 ppm. After 18 and 24 months no significantly increased values were found. The relative weight of the thyroid was not increased at ETU concentrations of 5 and 25 ppm, but starting from 125 ppm there was an increase in the relative weight of the thyroid from the 2nd test month to the 24th month. This increase was already seen at 125 ppm in the test groups that were kept for 2, 6 and 12 months, whereas in the tests that ran for 18 and 24 months this was only observed at 250 ppm and above. On histological examination at the end of the 6-months test, the thyroids of all male and female animals in the 500 ppm group were hyperplastic. Some of these thyroids exhibited adenomatous nodules and changes, which were regarded as carcinomas. The parathyroids of all animals were normal. Sections from the brain, Harder's gland, eye with retina and iris, extra-orbital lachrymal gland, salivary glands, salivary-gland lymph nodes, heart, lungs, liver, kidneys, gall bladder, ovaries, uterus, spleen, pancreas, adrenal glands, stomach, duodenum, jejunum, ileum colon, bone, bone marrow and skeletal muscle did not exhibit any changes that could be attributed to feeding of ETU. After 12 months, in all test groups and for both sexes starting from 5 ppm there was increased vascularization of the thyroid, and starting from 125 ppm a nodular hyperplasia was found in male rats; in female rats this was only seen at 500 ppm (see 7.7 regarding carcinogenicity). Also in the feeding tests lasting 18–24 months, hyperplasia of the thyroid was observed after administration of 5–250 ppm, being detected most frequently in the 25 and 125 ppm groups. In the two low-dose groups (5 and 25 ppm) the histological picture of hyperplasia was not very marked, but the thyroids showed marked macroscopic differences from the thyroids of the control animals, with regard to lobe formation, size of follicles, height of the glandular epithelium, colloid content, keratinization of the follicles and size of the organ (precise figures are not given). It could be noted that after 12–24 months of administration of 250 and 500 ppm, cataracts and keratitis were found in 6/69 and 12/70 rats respectively. However, such changes were also seen in the control group for 2/72 rats (Graham et al., 1973, 1975).

Another long-term test compared the effect of ETU on rat and hamster (breeds not stated). In each case 20 males and females of each species received ETU in the feed at concentrations of 0, 5, 17, 60 or 200 ppm for 24 months (rats) and 20 months (hamsters). Both species exhibited dose-dependent retarded growth and reduced food intake. In the case of rats, reduced weight gain was observed in the 60 and 200 ppm groups starting from the first month of testing. It also included the animals of the 17 ppm group starting from the second month. For the hamster this effect was only significant in the 60 and 200 ppm group. No sex differences were established for the rat or for the hamster. The reduced weight gain was also accompanied by reduced food intake, dependent on dose, and this was particularly evident in the 60 and 200 ppm groups. To check for hepatotoxic action of ETU, blood samples were taken at intervals of three months and the activities of alkaline phosphatase and glutamate-pyruvate transaminase (GPT) were determined. During checks after 3, 6 and 9 months, a temporary rise of the serum GPT values was found in the male and female rats in all dose groups and an increase in alkaline phosphatase activity was seen in male rats only, at all dose levels. After 12, 15, 18 and 24 months these changes could no longer be detected. In hamsters, GPT activity was unaffected by the treatment. During the first year of treatment the values for alkaline phosphatase activity were not significantly raised in either sex for any dose groups, but as the test continued they fell below the level in the control animals after 12 and 18 months. In both species and regardless of the sex of the animals, the cholesterol content of the serum was significantly ($p<0.01$) increased in all dose groups and at all points of time. With increasing ETU concentration, the cholesterol content of the serum also increased. Histopathological investigation of the rat liver showed no findings relating to feeding of ETU. An increased occurrence of haemorrhagic cysts was described in the thyroids starting from 60 ppm (no further details). In the hamster there was increased dose-dependent dystrophy of the gall ducts (Gak et al., 1976).

In a 30-week test, 10 male and 10 female Wistar rats (Morini breed) received 600 ppm ETU in the feed, with the aim of establishing which organs or functions, in addition to the thyroid, are damaged. 50% of the males and 20% of the females died within 80 days. After this time, all the rats survived to the end of the test. In the animals that survived, there was only a slight weight increase in comparison with the controls; all surviving rats exhibited alopecia (approx. 80%

hair loss) and pronounced conjunctivitis and blepharitis. The P 450 content in the liver was significantly lower shortly before the end of the test, whereas proteins, phospholipids and triglycerides were not significantly altered compared with the controls (Ugazio et al., 1985).

7.6 Genotoxicity

In vitro. In 17 independent studies with Salmonella typhimurium strains, ETU showed negative effects in 14 studies and positive effects in only 3 studies. In the last mentioned tests, a positive result was found once for strain TA 1535 (with and without S9-mix), in another laboratory for TA 98 (only with S9-mix) and in the third laboratory with TA 98 and TA 100 (only with S9-mix); in the latter laboratory, 3-methyl-cholanthrene-induced rat liver S9-mix was used for activation. When this experiment was repeated with phenobarbitone-induced rat liver S9-mix, ETU was no longer mutagenic. The results of the first two laboratories were reproducible (de Serres and Ashby, 1981, p. 129).

In another study, ETU was slightly mutagenic with strain TA 1530 without S9-mix (80 mg/plate, number of revertants increased by a factor of 2.1 compared with the controls) (Schupbach and Hummler, 1977).

In the case of strain TA 1950, ETU had a mutagenic effect at concentrations of 10 and 15 mg/plate without S9-mix (increase of number of revertants by a factor of 4), 20 mg/plate being toxic (Autio et al., 1982).

In another investigation with strains HisG 46, TA 1535, TA 1536, TA 1537, TA 1538 and TA 100, there was a dose-dependent increase in the number of revertants only for TA1535 without S9-mix from 5 mg/plate to 20 mg/plate; at 20 mg/plate by a maximum factor of 4 (Teramoto et al., 1977).

With Salmonella strain G 46, ETU was positive in a dose-related manner at concentrations of 10, 100 and 1000 ppm; at 10,000 and 20,000 ppm, ETU was toxic to bacteria (Seiler, 1974).

Without giving details, it was reported that ETU was weakly positive in the "Salmonella assay" (NTP, 1983).

In two studies with E. coli WP2 (WP2 B/r, WP2 uvrA, WP2 uvrA/pKM101), ETU did not have a mutagenic effect; in another test with E. coli K12/343/113, slight mutagenic activity was seen in a suspension test (de Serres and Ashby, 1981, p. 351, 385, 396).

With E. coli WP2 hcr$^+$ and WP2 hcr$^-$, ETU was not mutagenic at concentrations of up to 10 mg/plate (Teramoto et al., 1977).

In the host-mediated assay with strains G 46 or TA 1530 in Swiss albino mice when ETU was given i.p. at doses of 0, 500, 2000 and 6000 mg/kg (LD_{50} 5400 mg/kg), ETU was only weakly mutagenic (slight increase in the number of revertants for TA 1530 at 6000 mg/kg, factor 2.37; Schupbach and Hummler, 1977).

In another test with the strain HisG 46 (200 and 400 mg/kg oral, 3 times at intervals of 3 hours to each of 6 male ICL-SD rats or 6 female ICL-ICR mice), no increase in the numbers of revertants was observed (Teramoto et al., 1977).

Moreover, no increase in the number of revertants was found in strain TA 1950 following a single i.p. injection of 5 mg/kg ETU to male NMRI mice (test duration 3 hours; Autio et al., 1982).

In various DNA-repair tests, ETU gave different results. In the Rec assay with Bacillus subtilis (H17 Rec^+ and M45 rec^-), ETU was positive once (de Serred and Ashby, 1981, p. 175), but in a second investigation with the same strains it had no detectable effect (Teramoto et al., 1977).

In the Rec assay with Escherichia coli, no DNA damage was found (de Serres and Ashby, 1981, p. 183); on the other hand, in another Rec assay with Escherichia coli rec^-, ETU was positive in strains 2921, 9239, 8471, 5519 and 7623 with S9-mix (de Serres and Ashby, 1981, p. 193).

In a "Killing Test" with Escherichia coli WP2,WP67 and CM871, ETU showed no effect with and without S9-mix (de Serres and Ashby,1981, p. 199).

In a similar study (suspension test) with Escherichia coli W3110 ($polA^+$) and Escherichia coli P3478 ($polA^-$), a weak mutagenic effect was found only with S9-mix (de Serres and Ashby, 1981, p. 210).

In these six DNA-damage-repair studies just mentioned, 3 had a positive result and 3 a negative result. No clear conclusions can be drawn, and the results must be regarded as equivocal. Of 8 studies reported in the literature investigating mutagenic activity in yeasts, 4 gave positive results and the other 4 gave negative results (de Serres and Ashby, 1981, p. 414, 424, 434, 457, 468, 481, 491, 502). Different test systems and yeast strains were employed in all 8 test procedures. Therefore it is not possible to draw conclusions about the mutagenicity of ETU to yeasts, and the results are to be regarded as equivocal.

In tests in the UDS system (unscheduled DNA synthesis) on human fibroblasts (WI-38 cells), ETU proved to be negative in 2 separate studies (de Serres and Ashby, 1981, p. 517, 528), whereas in HELA S3 cells, it was weakly positive in the UDS test without S9-mix (de Serres and Ashby, 1971, p. 83, 533). No induction of SCE (sister chromatid exchanges) in CHO (Chinese hamster ovary) cells was found in 3 independent tests (de Serres and Ashby, 1981, p. 538, 551, 560).

According to investigations conducted in the NTP programme, again no SCE's were triggered by ETU (no further details; NTP, 1984).

In tests for chromosome aberrations in vitro on CHO cells, no effects were found up to 5 mg ETU/ml (de Serres and Ashby, 1981, p. 551), and in another study no definite results were obtained up to 200 µg/ml (de Serres and Ashby, 1981, p. 570).

According to reports from the NTP, no chromosome aberrations were triggered by ETU in CHO cells (NTP, 1984).

ETU was also negative in another chromosome aberra- tion test on hamster cells (Teramoto et al., 1977).

In tests of ETU on mouse lymphoma cells (L5178Y) and in the induction of mutations at multiple genetic loci of CHO cells there were no indications of a mutagenic effect (de Serres and Ashby, 1981, p. 580, 594).

In the cell transformation test on baby-hamster kidney cells (BHK21) without metabolic activation, a positive effect of ETU was found in 2 independent investigations (de Serres and Ashby, 1981, p. 626, 638).

In vivo. In 3 different tests with Drosophila (sex-linked recessive lethal mutations) no mutagenic activity was demonstrated (de Serres and Ashby, 1981, p. 651; Mollet, 1975; Woodruff et al., 1985), and questionable results were reported in another investigation (NTP, 1985).

ETU did not induce lethal mutations in 2 different studies on mice (1st publication: 0, 500, 1000 and 3500 mg/kg once orally, mating time 10 weeks; 2nd publication: 0, 300 and 600 mg/kg orally on 5 successive days, mating time 5 weeks; Teramoto et al., 1977; Schupbach and Hummler, 1977).

No SCE's were triggered by ETU in in vivo investigations on male CBA/l mice (de Serres and Ashby, 1981, p. 673).

A chromosome analysis in vivo on male and female Wistar rats did not show any chromosomal changes in the bone marrow cells (doses up to 400 mg/kg, twice with a 24-hr interval, oral; Teramoto et al., 1977).

The reported 3 micronucleus tests on mice (80% of LD_{50} i.p. twice with a 24-hour interval, or once up to 880 mg/kg i.p. (LD_{50} 1770 mg/kg)) were negative (de Serres and Ashby, 1981, p. 686, 698).

Another test of ETU in this test system (700, 1850 and 6000 mg/kg ETU twice orally within 24 hours) also did not show any increase of micronuclei in the polychromatic erythrocytes (Schupbach and Hummler, 1977).

An i.p. injection on 5 successive days in male B6C3F1 mice at doses of 166, 332, 664, 1328 and 6 255 mg/kg did not produce any morphological abnormalities of the spermatozoa (de Serres and Ashby, 1981, p. 712).

A single i.p. injection of 100 mg/kg did not lead to any inhibition of testicular DNA synthesis (Seiler, 1977).

The data on the mutagenicity of ETU are contradictory. In some studies ETU, mostly only at high concentrations, induced mutations in bacteria and fungi; in addition there are indications of a possible DNA-damaging effect of ETU in prokaryontes, yeasts and mammalian cells in vitro with and without metabolic activation. However, a roughly equal number of investigations using these test systems gave negative results. Even the results of the available UDS tests are not uniform. In animal cell cultures, no indications were found of a genotoxic effect (sister chromatid exchange and chromosome aberrations). However, ETU is positive in the cell transformation test without metabolic activation. No indications of a mutagenic effect are found in the host-mediated assay (mouse, rat) or in investigations on Drosophila melanogaster (sex-linked recessive lethal mutations). No dominant lethal mutations, no micronuclei, no sister chromatid exchanges, and no morphological abnormalities of the spermatozoa or indications of inhibition of testicular DNA synthesis were found in tests on mice. ETU does not induce any chromosome aberrations in rats.

7.7 Carcinogenicity

Mice of breeds C57BL/6xC3H/Anf-F1 (breed X) and C57BL/6xAKR-F1 (breed Y) received 215 mg/kg ETU daily by stomach tube from the 7th to the 28th day of life, and then 646 ppm ETU in the feed for 72 weeks. 18 mice per breed and per sex were used. There were several negative control groups, which also served as controls for other tests in progress simultaneously. The animals were dissected when they died, at the latest 82 weeks after the beginning of the test. Hepatomas, lung tumours and lymphomas were assessed.

Table 3. Tumour incidence after oral administration of ETU in mice

	Breed	Sex	Total number mice dissected	Mice with hepatomas	Mice with lung tumors	Mice with lymphomas
Control	X	Male	79	8	5	5
Control	X	Female	87	0	3	4
Control	Y	Male	90	5	10	1
Control	Y	Female	82	1	3	4
ETU	X	Male	16	14	1	0
ETU	X	Female	18	18	3	1
ETU	Y	Male	18	18	2	3
ETU	Y	Female	16	9	0	4

The types of tumours were not further classified. The results are summarized in table 3.

According to these investigations, ETU led only to a significant incidence of liver tumours in the two breeds of mice (Innes et al., 1969).

Male and female Charles-River rats (68 male and 68 female animals/group, 5 weeks old) received 0, 5, 25, 125, 250 or 500 ppm ETU in the feed for 12 and 24 months. After 66 weeks, 3 males and 3 females were taken from each test group and were kept on a diet without ETU for up to 2 years. The tumour yield from this test is shown in the following table 4 (there was no differentiation between male and female rats):

Table 4. Pathological changes in the thyroid of ETU-treated male and female Charles-River rats

	Dose (ppm ETU in feed)					
	0	5	25	125	250	500
Male and female rats	136	136	136	136	136	136
Rats surviving after 1 year	72	75	73	73	69	70
Rats surviving after 2 years	32	39	43	37	32	18
Thyroid carcinoma	2	2	1	2	16	62
Thyroid adenoma	2	0	5	1	21	3
Thyroid hyperplasia	4	20	41	44	27	3
Hepatoma	1	1	1	2	1	5

According to these investigations, ETU caused carcinomas and adenomas of the thyroid starting from 250 ppm in the feed, as a function of dose. There was no increase in the incidence of thyroid tumours up to 125 ppm. At 5 ppm and above, hyperplasia of the thyroid was seen, its intensity increasing in relation to the dose. The hyperplasia in the rats which were put on a normal diet after 66 weeks until the end of the 2-year test was not substantially reversible (Graham et al., 1975).

Another carcinogenicity test was conducted with Charles-River CD rats given ETU in the diet. The maximum tolerable dose in the feed (355 ppm) was determined first, in a preliminary test. The following doses were used in the carcinogenicity test: 0, 175 and 350 ppm ETU in the feed. The period of administration was 78 weeks, and the total test duration was 104 weeks. 26 male and 26 female newly-weaned rats were used per dose. Moribund animals and animals with clearly palpable neoplasms were killed. The survival rate of the animals to the 78th week was control 86% (male) and 84% (female), in all treated groups 89%; to the 104th week, control 53% (male) and 46% (female), in all treated groups 60%. The investigation produced the following results:

Table 5. Thyroid carcinomas after 2 years in Charles-River CD rats treated with ETU (total test duration 24 months, substance administrated for 18 months) (Ulland et al., 1972; Weisburger et al., 1981)

	Control	175 ppm	350 ppm
Rat (male)	0/30	2/26 (12%)[a]	17/26 (66%)[c,e]
Rat (female)	0/30	2/26 (12%)[b]	8/26 (31%)[d]

[a] 1st tumour in the 104th week
[b] 1st tumour in the 91st week
[c] 1st tumour in the 78th week
[d] 1st tumour in the 68th week
[e] 2 animals with lung metastases

In all treated animals of both test groups there was diffuse hyperplasia of the thyroids (Ulland et al., 1972; Weisburger et al., 1981).

Groups of 20 male and 20 female rats or hamsters (breed not stated for either species) received ETU in the feed for 24 months and 20 months, respectively, in concentrations of 0, 5, 17, 60 and 200 ppm. In rats, there was a dose-related incidence of tumours (% tumours in the animals used, no precise details) as shown in table 6:

Table 6. Tumour incidence after oral ETU administration in rats

	0 ppm	5 ppm	17 ppm	60 ppm	200 ppm
male	0	0	5.88	42.1[a]	82.4[b]
female	5.3	6.3	18.8	22.2	56.3[b]

[a] $P = 0.01$
[b] $P = 0.001$

Essentially these were thyroid tumours, although in male rats testicular tumours were found, originating in the Leydig cells, and in female rats 2 thymomas were also found (no precise details). In hamsters the incidence of thyroid tumours was not increased at the administered doses (no details; Gak et al., 1976).

In these 3 feeding tests on rats, ETU induced follicular carcinomas of the thyroid, as a function of dose. Both sexes were affected, but the males exhibited a higher incidence. Even at 5 ppm thyroid hyperplasia was seen, and this increased in a dose-related manner. There was minimal regression after ETU was discontinued. Liver tumours developed in mice. ETU did not induce any tumours in a feeding study with hamsters.

7.8 Reproductive toxicity

The individual data from the literature (Stula and Krauss, 1977; Lu and Staples, 1978; Khera, 1973; Teramoto et al., 1978a, b, 1981; Ruddik and Khera, 1975; Chernoff et al., 1979; Khera et al., 1983; Hung et al., 1986; Lu et al., 1980) are presented in table 7 "Embryotoxic and teratogenic effects of ETU". The findings recorded in these studies are summarized below.

Rat. ETU was teratogenic in rats at oral doses which did not produce any detectable maternal or foetal toxicity (Stula and Krauss, 1977; Lu and Staples, 1978; Khera, 1973; Teramoto et al., 1981; Ruddik and Khera, 1975; Chernoff et al., 1979; Hung et al., 1986). At higher doses, foetal toxicity was also observed (Teramoto et al., 1978a, 1981; Plasterer et al., 1985). The observed malformations occurred not only with repeated administration of doses of 40, 20, 10 and 5 mg/kg on days 6 to 15 and 7 to 21 of pregnancy, but also with administration on individual days (single doses 60, 120, 240 and 250 mg/kg orally), a teratogenic effect being seen from the 10th to the 20th day of pregnancy (period of organogenesis; Teramoto et al., 1978b, 1981; Ruddik and Khera, 1975; Hung et al., 1986). With single administrations before the 10th day (days 6 to 9), no teratogenic effects were observed even at high doses (Ruddik and Khera, 1975).

The following dose-dependent malformations were found:
- CNS malformations (hydranencephaly, hypoplastic cerebellum, over-distention of the ventricles, hydrocephalus internus, plus exencephaly and meningocele)
- skeletal malformations (kinked tail, oligodactylia, syndactylia, brachygnathia, micromelia, shortened and fused ribs, cleft palates, missing bones in the extremities)
- kidney malformation (with single administration of high doses, e.g. 250 mg/kg, hydronephrosis and hypoplastic kidneys)
- in additon, anal atresia and hypoplasia of the genitals were found.

On repeated administration, the threshold dose was 5 and <5 mg/kg (Khera, 1973; Chernoff et al., 1979). ETU also had a teratogenic effect when applied to rat skin (on days 11 and 12 or 12 and 13 of pregnancy, 25 and 50 mg/kg, once daily in DMSO). 25 mg/kg caused no malformations, 50 mg/kg caused encephalocele and various skeletal malformations (Stula and Krauss, 1977).

Mouse. With regard to teratogenic action of ETU, mice were less sensitive than rats and hamsters. On repeated oral administration of 100 and 200 mg/kg to CD1 mice on days 6 to 10 of pregnancy supernumerary ribs were seen at 200 mg/kg (100 mg/kg: no findings; Chernoff et al., 1979). In another investigation (7th to 15th day of pregnancy, ICL-ICR mice) no malformations were seen at 200, 400 and 800 mg/kg (Teramoto et al., 1978a). A single oral dose of 2000 mg/kg to ICR mice on the 10th day of pregnancy led to a slight increase in the percentage of malformed foetuses (3.2% foetuses malformed, against 0.9% in the controls). Only skeletal malformations were found (Teramoto et al., 1981).

Rabbit. On oral administration of 0, 10, 20, 40 and 80 mg/kg to white New Zealand rabbits from days 7 to 20 of pregnancy, increased resorptions only occurred at the highest dose of 80 mg/kg, with a reduction in brain weight and degeneration of the proximal tubules, but no malformations. The other doses caused no changes in the foetuses (Khera, 1973).

Hamsters. Single oral administration (11th day of pregnancy) of high doses (600 to 2400 mg/kg) led to malformations of the CNS and skeleton (Khera et al., 1983). Repeated oral administration of ETU to golden hamsters on days 5 to 10 or 6 to 13 of pregnancy did not cause malformations up to a dose of 100 and 90 mg/kg respectively (Teramoto et al., 1978a; Chernoff et al., 1979). At 270 mg/kg, up to 14% vertebral anomalies were observed and at 810 mg/kg, in addition to foetotoxic effects (reduced number of live foetuses, significantly reduced foetal body weight), malformations, mainly of the skeletal system, were observed in 63% of the foetuses examined (Teramoto et al., 1978a). Repeated oral administration of 120 to 360 mg/kg on days 7 to 14, 7 to 11 or 11 to 12 of pregnancy cause malformations of the skeleton and CNS. At 360 mg/kg, aplasia of the kidneys was also seen (Lu et al., 1980).

Guinea-pigs. ETU did not caused any malformation when administered at oral doses of 0, 50 and 100 mg/kg on days 7 to 25 of pregnancy (Chernoff et al., 1979).

Cats. ETU was administered orally to pregnant cats on days 16 to 35 of pregnancy, at doses of 0 (14 animals), 5 (9 animals) 10 (7 animals), 30 (8 animals), 60 (7 animals) and 120 (10 animals) mg ETU/kg. Doses of 10 to 120 mg/kg were toxic to the mothers (weight loss, ataxia, paralysis of the hind quarters). In the cats of these dose groups that died or were killed when moribund, there were some cases of malformations (malformation of the skeleton, coloboma, umbilical hernia). 5 mg/kg was tolerated by the mothers without toxic symptoms; of the 4 foetuses available for investigation from this dose group, 1 foetus exhibited encephaly, hydrocephaly and cleft palate. No definite statement can be made concerning teratogenic action of ETU in cats on the basis of these investigations (Khera and Iverson, 1978).

According to the aforementioned investigations, the rat is especially sensitive to ETU. ETU has teratogenic effects if administered during the period of organogenesis at doses that are not toxic to the mother. A teratogenic effect of ETU can also be detected in the hamster, though at higher doses, reaching maternaly-toxic levels. The mouse is less sensitive to ETU than the rat or hamster, but malformations also occur in this species at high doses. No definite statement can be made concerning the teratogenic action of ETU in guinea-pigs, rabbits and cats.

To determine the reproductive index, 600 mg/kg (MTD, LD_{50} 5085 mg/kg) was administered orally to 35 mice on days 7 to 14 of pregnancy. This dose caused no maternal toxicity and no weight loss in comparison with the untreated controls. Of the 35 surviving mothers, 26 gave birth. Therefore the reproductive index was lowered to 0.734 ($p<0.05$). The mortality rate of the offspring per litter and the average weight of the newborn pups in the first 3 days were unaffected (Plasterer et al., 1985).

ETU was also tested on chicken embryos. ETU was applied intracardially to 3-day-old chicken embryos (30 embryos per dose) at doses of 1, 2, 3 µmol/egg. The embryos were examined for malformations on the 14th day of incubation. The percentage of malformations observed during incubation and at the end of the test was 2, 13, 13 and 41% at 0, 1, 2 and 3 µmol/egg respectively (eye malformations, open coelum, defects on the back and the nape, wing malformations; Korhonen et al., 1982).

According to findings that were only reported in tables, relating to a "Short term in vivo reproductive toxicity assay" (no further details) using doses of 100, 200, 300 and 600 mg/kg (route of administration not stated), at 300 mg/kg there was a decrease in the number of young still alive after 12 hours, and at 600 mg/kg there was a decrease in the number of litters with one or more young. Increased maternal mortality was not observed at any of the doses (NTP, 1984).

For histological investigation of the neural canal and of the front limb buds, rats were given 100 mg/kg on the 12th day of pregnancy or 200mg/kg ETU on the 13th day of pregnancy. Three animals were killed 3, 6, 12, 24, 48, 72 and 96 hours after treatment and corresponding controls were also conducted. It was known (see above) that there was a high incidence of malformations (central nervous system and limbs) at these doses. At a dose of 100mg/kg on the 12th day of pregnancy, necrotic cells were found in the wall of the neural canal of the foetuses after just 12 hours. After 24 hours there was severe cell necrosis, the ventricles were expanded and filled with cell residues, and brain development was delayed. At 200 mg/kg on the 13th day of pregnancy, brain development was severely delayed, the ventricles tended to a uniform large cavity. At 100 mg/kg on the 12thday of pregnancy there were no histological changes in the front limb buds, whilst at 200 mg/kg on the 13th day of pregnancy they showed progressive necrosis (Teramoto et al., 1978b).

When 30 mg/kg was given orally to rats on the 13th day of pregnancy, first there was karyorrhexis of the neuroblasts at the basal lamina of the CNS. Lesions were seen 12 hours after administration, leading to the development of hydrocephalus in the animals after birth. At a dose of 5 mg/kg, only occasional damage to the neuroblasts were seen; hydrocephalus did not develop in subsequent breeding (Khera and Tryphonas, 1985).

The neuroblast therefore appeared to be the specific target for ETU. This result was supported by in vitro and in vivo investigations. In cell cultures (mixture of neuronal and non-neuronal gliocytes, obtained from foetal brain), ETU (≥ 0.5 mM) caused necrosis of the neuronal cells and definite depression in the formation of neurons and fascicles, without altering the non-neuronal cells. In a cell suspension of foetal brain from ETU-treated mothers the cell count was also noticeably lower than for corresponding controls. When this cell suspension was cultured in "mono-cell layers", there were greatly reduced populations of neuronal cells, whereas the non-neuronal cells were unaffected or were increased (Khera, 1986).

In pregnant rats (13th day of pregnancy), simultaneous administration of ETU (60 mg/kg, oral) and SKF-525A (2-diethylaminoethyl-2,2-diphenylvalerate hydrochloride (40 mg/kg i.p.)) and ETU and NMI (N-methyl-2-thio-imidazole (200 mg/kg, oral)) – SKF-525A and NMI are considered to be inhibitors of metabolism – caused an increase in the incidence and severity of foetal malformations in comparison with ETU alone. It could be assumed on the basis of these results that ETU itself, and not a metabolite, is to be regarded as the proximal teratogen (Khera and Iverson, 1981).

The following studies are available concerning post-natal toxicity of ETU: Pregnant Wistar rats (5–10/group) received 0, 15, 30 or 45 mg/kg ETU orally on the 15th day of pregnancy. The progeny of the 15 mg/kg group developed without clinical abnormalities. In the 30 mg/kg group, 90% of the offspring exhibited hydrocephalus and microphthalmia. The animals affected died within 9 weeks post partum. The surviving offspring exhibited motor disturbances in the extremities and a jerky gait. The cerebral hemispheres consisted of thin-walled, fluid-filled cavities. Histologically, there were lesions to the neuroaxis. The offspring from the 45 mg/kg group all died within 4 weeks. 90% of them had hydrocephalus and microphthalmia, and histologically there was necrosis of the axis cylinders of the nerves. No such changes were observed in the offspring of the untreated animals (Khera and Tryphonas, 1977).

Pregnant Sprague-Dawley rats (6–28 animals/group) received 0, 20, 25 or 30 mg/kg ETU orally on days 7 to15 of pregnancy. After birth, the litters were investigated for the following parameters for up to 43 days: weight, opening of the eyes, startle reflex from noises and righting reflex when dropped from a height of 25 cm. For the animals in the 20 and 25 mg/kg group, these parameters were not significantly different from the control group. In the 30 mg/kg group, 6 litters (55%) died 1 to 2 days after birth. The animals did not drink, although the mothers had sufficient milk. Of the surviving young from this group, 40% exhibited hydrocephalus. There were no changes in the animals in the other groups (Chernoff et al., 1979).

According to another investigation on CD rats involving oral administration of 30 mg/kg on days 7 to 15 of pregnancy, 4 out of 9 litters died in the first 24 hours after birth. There were no morphological defects. Only 2 litters survived the first week after birth. In these animals there was formation of hydrocephali. 20 mg/kg was tolerated, apart from hyperactivity of the newborn animals (Kavlok and Chernoff,1978).

In another test, pregnant Wistar rats (number not stated) received a single oral dose of 30 or 50 mg/kg ETU on the 18th, 19th or 20th day of pregnancy, 10 or 20 mg/kg on the 17th, 18th, 19th or 20th day or 1, 3 or 5 mg/kg on the 18th day of pregnancy. Viability, weight gain, behaviour and visible defects of the offspring were recorded. Administration of 1, 3 or 5 mg/kg on the 18th day did not produce lesions in the offspring. The number of stillbirths was significantly increased on administration of 30mg/kg (16%) and 50 mg/kg (46%) on the 20th day. At 50 mg/kg on the 20th day, the animals had significantly lower birth weights; all the offspring of this group had hydrocephalus and died within 4 days of birth. Out of 107 offspring of the 20 mg/kg group (treated on the 18th, 19th or 20th day), which were killed after 6 months, 28 had hydrocephalus. Out of 119 offspring of the 10 mg/kg group, there were 19 rats with hydrocephalus after 6 months. Weight gain during the suckling period was the same in all the test groups was in the control group, in the animals that survived (Lewerenz and Bleyl, 1980).

In single application on the 11th, 12th, 13th and 14th day of pregnancy, short tails or absence of tails was seen at all doses, and in addition omphaloceles, gastroschisis, syn- and polydactylism were seen, but were not analysed in detail.

7.9 Effects on the immune system
No information available.

7.10 Other effects
The hepatic monooxygenases were affect-ed by ETU differently in rats and mice. After oral administration for 3 days of 50 and 75 mg/kg to rats and 50, 75, 100, 500 and 1000 mg/kg to mice, there

was a decrease of aminopyrine-N-demethylase in rats to 60–70% of the controls (24 hours after treatment) and decreases in aniline-hydroxylase activity and P 450 content on the 3rd day after treatment. In the mice there was an increase in the P 450 content at all doses and increased aniline-hydroxylase activity at 100 mg/kg and higher. Aminopyrine-N-demethylase was unaffect-ed. In the authors' opinion, the different response of the hepatic monooxygenases could be partially responsible for the differences in toxicity and teratogenicity of ETU in rat and mouse (Iverson et al., 1980).

200 mg/kg was administered i.p. to male Wistar rats (160–220 g) on 4 consecutive days. Investigations of the Nervus ischiadicus in the rats thus treated showed only slight changes in the myelin structure in light-microscopic and electron-microscopic preparations, and it was presumed that these changes were caused by degenerative processes. In immunofluorescence studies conducted with the same substrate, only slight perinuclear accumulation of filaments was found in comparison with the controls (Lehto et al., 1984).

8. Experience in humans

In a 53-year-old woman who had worked in a factory producing rubber goods for 13 years, an itching eruption developed on the hands and arms. There was improvement at weekends, and the situation became worse when work began. A standard patch test (ICDRD) as performed. The woman reacted positively to nickel and cobalt and to rubber material, which she worked with. On testing the individual constituents of this rubber, there was a positive reaction to ETU. A positive reaction could be established up to a concentration of 0.01%, but it was negative at 0.001%. 20 control subjects reacted negatively with 1% ETU solution (Bruze and Fregert, 1983).

In 2 factories in which ETU was handled in different ways, workers were investigated clinically and with regard to thyroid function over a period of 3 years at invervals of about 6 months. In the first factory (manufacture of ETU) the average length of service of the10 workers employed in ETU manufacture was 10 years (5–20). Concentration measurements were made in 1976. The ETU dust concentration was between 10 and 240 $\mu g/m^3$, and in measurements with the personal air sampler it was up to 330 $\mu g/m^3$. Up until 1977, not all workers wore protective clothing, and the ventilation of the rooms was unsatisfactory. After this the protective equipment was improved, but concentration measurements were not repeated. Then a total of 23 measurements of T4 were made for these workers (age 26–62 years). The control group consisted of 16 subjects. The studies were undertaken in the years 1979–1982 (no exact details available). In the 2nd factory, 5 workers (so-called mixers, in the age range 28–56 years, total of 22 measurements of T4) were investigated, who had been exposed to concentrations of 120–160 µg/m^3 (personal sampler) since 1977. There were 10 control subjects. Measurements of thyroxin (T4) were made at 6-month intervals in the years 1979–1982). None of the workers had a thyroid disease or was taking associated medication. The mean thyroxin values (T4) were lower in the exposed workers who were employed in the manufacture of ETU (1st factory), with greater variation than in the controls (all controls 105.7 ±0.0054 nmol/l, 1st factory 96.4 ± 0.02 nmol/l, p>0.10). In the mixers (5 workers, 2nd factory) the mean values for T4 were significantly (p<0.01) lower than in the controls (80.5 ± 0.0012 nmol/l; Smith, 1984).

In another retrospective study, 1929 workers (sex not stated) from various factories in Birmingham who had been in contact with ETU in rubber manufacture since 1952 were investigated for the occurrence of thyroid cancer. No case of thyroid cancer occurrred in this group. The cancer register for the town of Birmingham recorded 49 cases of thyroid cancer in the general population in the period from 1957 to 1971. No further details are given concerning the group investigated or concentration measurements (Smith, 1976).

In a rubber-manufacturing factory in Birmingham, where ETU was used, all women of child-bearing age were asked in a retrospective study about malformations in their children. Out of 699 women, born in 1918 or later, who had left the factory between 1963 and 1971, 255 women had produced a total of 420 children. Of the women questioned, only 59 women had worked in the factory during early pregnancy. Out of the total of 420 children, 11 had malformations. The observed number of malformations in this group was no higher than the expected number in the general population of the town of Birmingham. No details were given concerning abortions or concentration measurements (Smith, 1976).

Table 7. Embryotoxic and teratogenic effects of ETU

Animal species and breed	Number per group	Dose and route of administration	Day of pregnancy		Maternal toxicity	Findings in foetuses	References
			Treatment	Sacrificed			
Rat, Wistar	5–10	240 mg/kg single oral	6, 7, 8, 9	22	None	No malformations	Ruddik and Khera, 1975
			10	22	None	Malformations of external genitalia (6%) Hydronephrosis or hypoplastic kidneys (20%) Tail abnormalities (20%)	
			11	22	None	Spina bifida (30%) Fused ribs (100%) Kidney abnormalities (75%) Tail abnormalities (100%) Increased number of resorbed embryos (13%, control 8.5%)	
			12, 13, 14, 15	22	None	Foetal weight less than in control group hydrocephalus (100%) Exencephaly (90%) Brachygnathia (58–74%) Hydronephrosis (100%) Tail missing (67–100%) Micromelia of front extremities (90–100%)	
			16	22	None	Hydranencephaly (100%) Cleft palate (100%) Hydronephrosis (100%)	

Species		Dose	Treatment period	n	Mortality	Effects	Reference
	17, 18, 19			22	None	Hydranencephaly (100%) Hydronephrosis (100%) Hypoplastic kidneys (100%)	
	20, 21			22	None	Kidney malformations (20–42%)	Khera, 1973
Rat, Wistar	10–17	0, 5, 10, 20, 40, 80 mg/kg daily oral	21 to 42 days before mating up to 15th day of pregnancy	22	None	40 mg/kg: Malformed foetuses (98%) Exencephaly (24%) Micrognathia (19%) Kinked tail (57%) Delayed ossification (52%) Over-distension of ventricles (84%) Hypoplastic cerebellum (98%) 20 mg/kg: Malformed foetuses (69%) Over-distension of ventricles (45%) Hypoplastic cerebellum (64%) 10 and 5 mg/kg: Delayed ossification (42 and 33%)	
	6–15			22	80 mg/kg/day: lethal in 9/11 animals after 7–8 days	80 mg/kg: Malformed foetuses (100%) Micrognathia (75%) Coloboma of eyelids (58%) Oligodactylia (58%) Shortened tail (80%) Delayed ossification (100%) Exencephaly (100%) Meningocele (100%)	

Table 7 (continued)

Animal species and breed	Number per group	Dose and route of administration	Day of pregnancy Treatment	Day of pregnancy Sacrificed	Maternal toxicity	Findings in foetuses	References
Rat, Wistar	10–17	0, 5, 10, 20, 40 mg/kg daily oral	6–15 days before mating up to 15th day of pregnancy	22	40 mg/kg none	Over-distension of ventricles (100%) Hypoplastic cerebellum (100%) 40 mg/kg: Malformed foetuses (95%) Micrognathia (20%) Shortened tail (33%) Kinked tail (42%) Delayed ossification (67%) Exencephaly (21%) Meningocele (21%) Over-distension of ventricles (95%) Hypoplastic cerebellum (70%) 20, 10, 5 mg/kg: Delayed ossification (9–20%)	Khera, 1973
			7–20	22	None	40 mg/kg: Malformed foetuses (100%) Micrognathia (7%) Shortened tail (43%) Kinked tail (34%) Hydrocephalus (34%) Meningocele (32%) Over-distension of ventricles (100%) 20 mg/kg: Malformed foetuses (41%)	Khera, 1973

Ethylene thiourea 115

Species	Days of gestation	Route/dose			Findings	Reference
Rat, ChR-CD Sprague-Dawley	5–8	Percutaneous on skin of back, in each case 25 and 50 mg/kg in DMSO once daily	10, 11		Kinked tail (3%) Hydrocephalus (3%) Over-distension of ventricles (19%) Hypoplastic cerebellum (41%) 10 mg/kg: Malformed foetuses (2%) Over-distension of ventricles (1%) Hypoplastic cerebellum (2%) 5 mg/kg: no findings	Stula and Krauss, 1977
			12, 13	20	No data	
				20	25 mg/kg: No malformations 50 mg/kg: Shortening of tail (3/83 foetuses) Fused ribs (2/83) 50 mg/kg: All 73 foetuses malformed: encephalocele, tail missing, missing bones in extremities, shortened lower jaw, fusing of ribs and sternum	Stula and Krauss, 1977
Rat, Charles-River	6–12	40 mg/kg oral, daily	7–15	20	No data	Lu and Staples, 1978
					Malformations in 121/142 foetuses: Craniocele (78) Meningorrhoea (10) Ectrodactylia (12) Talipes (50) Shortened tail (40) Kinked tail (46) Hydrocephalus internus (75)	

Table 7 (continued)

Animal species and breed	Number per group	Dose and route of administration	Day of pregnancy - Treatment	Day of pregnancy - Sacrificed	Maternal toxicity	Findings in foetuses	References
						Fused ribs (2) Shortened ribs (4) Control: no malformations observed; simultaneous treatment with ETU and thyroxin malformations as with ETU alone (no impairment through altered thyroxin level)	
Rat, Wistar Imamichi	5–8	50, 100, 200 mg/kg oral	12 or 13	20	No data	12th day: Malformations, meningocele, short tail, oligo-, syndactylia 50 mg/kg (89%) 100 mg/kg (100%) 200 mg/kg (100%) Starting from 100 mg/kg necrosis of neuroaxis, at 200 mg/kg necrosis of the mesenchymal cells in the front paws 13th day: Malformations, meningocele, cleft palate, genital, hyperplasia, oligo-, syndactylia 50 mg/kg (0%) 100 mg/kg (100%) 200 mg/kg (100%) Starting from 100 mg/kg necrosis of neuroaxis	Teramoto et al., 1978 b

Rat, Wistar	5–6	250 mg/kg once oral (LD 50 ca 1000 mg/kg)	12 or 14	20	No overt toxic symptoms	Foetal weight significantly lower than for the controls (3.67 g) with application on 12th day: 2.36 g with application on 14th day: 1.89 g All foetuses malformed after application on 12th or 14th day: Meningocele (100%) Micrognathia (100%) Cleft palate (100%) Omphalocele (100%) Shortened or kinked tail (100%) Anal atresia (97%) Hypoplasia of genitals (100%) Oligodactylia (100%) Bony ankyloses on sternum and vertebrae (51 and 44% resp.) Bones missing in extremities (34–51%) Microphthalmia (100%) Hypoplasia of bulbus olfactorius (100%) Pulmonary hypoplasia (100%) Hydronephrosis (53%)	Teramoto et al., 1981
Rat, Sprague-Dawley	8–31	0, 5, 10, 20, 30, 40, 80 mg/kg oral, daily	7–21	21	At 80 mg/kg 25% mortality, weight decrease	80 mg/kg: Reduced birth weight (2.6 g, control 4.6 g) reduced number of sternal and caudal ossification centres, hydrocephalus, encephalocele, cleft palate, kypho-	Chernoff, et al., 1979

Table 7 (continued)

Animal species and breed	Number per group	Dose and route of administration	Day of pregnancy Treatment	Day of pregnancy Sacrificed	Maternal toxicity	Findings in foetuses	References
					40 mg/kg none	sis, mircromelia, oligodactylia, syndactylia, micrognathia 40 mg/kg: Reduced birth weight, reduced number of ossification centres, hydrocephalus, encephalocele, cleft palate	
					30 mg/kg none	30 mg/kg: Hydrocephalus, fewer caudal ossification centres	
					20 mg/kg none	20 mg/kg: Hydrocephalus	
					10 mg/kg none	10 mg/kg: No findings	
					5 mg/kg none	5 mg/kg: No findings	
Mouse, CD-1	25	0, 100 and 200 mg/kg oral daily	7–16	18	Increased liver/body weight ratio at 100 and 200 mg/kg ($p<0.01$)	200 mg/kg: Supernumerary ribs (31, control 17, $p<0.05$) 100 mg/kg: No findings	Chernoff et al., 1979
Golden hamster (Ela-ENG)	9–15	25, 50, 100 mg/kg oral daily	5–10	15	None	No malformations observed	Chernoff et al., 1979

Species	Days	Dose	n	?	Findings	Reference
Guinea pig, Hartley	3–6	0, 50 and 100 mg/kg oral	7–25	35	None	No malformations observed	Chernoff et al., 1979
Rabbit, white New Zealander	5–7	0, 10, 20, 40, 80 mg/kg oral	7–20	30	None	80 mg/kg: Indication of kidney damage (degeneration of proximal tubules, no further details) Increased resorptions and reduced brain weight in foetuses 40, 20, 10 mg/kg: No findings	Khera, 1973
Mouse, ICR	12	2000 mg/kg oral (LD 50 >2000 mg/kg)	10	18	None	3.2% of foetuses malformed (control 0.9%) Cleft palate (2.5%) (control 0%) Defects of cervical vertebrae (12.5%) (control 5.8%) Defects of caudal vertebrae (7.5%) (control 0%)	Teramoto et al., 1981
Syrian golden hamster	9–10	0, 600, 1200, 1800, 2400 mg/kg once, oral	11	15	None	Control: Defective sternum (41%) Increased number of ribs (24%) 600 mg/kg: Defective sternum (42%) 1200 mg/kg: Cleft palate (17%) Hydrocephalus (11%) Hypoplastic cerebellum (41%) Defective sternum (63%) 1800 mg/kg: Cleft palate (60%) Toe abnormalities (6%) Hydrocephalus (24%)	Khera et al., 1983

Table 7 (continued)

Animal species and breed	Number per group	Dose and route of administration	Day of pregnancy Treatment	Day of pregnancy Sacrificed	Maternal toxicity	Findings in foetuses	References
						Hypoplastic cerebellum (68%) Delayed ossification (19%) Defective sternum (84%) 2400 mg/kg: Resorption (30%) (control 0%) Dead foetuses (24%) (control 0%) Cleft palate (85%) Toe abnormalities (13%) Hydrocephalus (61%) Hypoplastic cerebellum (72%) Delayed ossification (51%) Defective sternum (86%)	
Rat, Wistar	8–10	0, 10, 20, 30, 40, 50 mg/kg oral	6–15	20	None	From 30 mg/kg: Foetal weight significantly reduced From 10 mg/kg: Dose-dependent increase in over-distension of lateral ventricle 10 mg/kg: 2% 20 mg/kg: 39% From 30 mg/kg: 100% 30 mg/kg: shortened and kinked tails (79%) 40 mg/kg: shortened and kinked tails (100%)	Teramoto et al., 1978a

Ethylene thiourea

Species	Dose	Days	(col)	Findings	Reference
Mouse ICL-ICR	0, 200, 400, 800 mg/kg oral	6–14	7–15	18	None	Meningocele (66%) 50 mg/kg: Shortened and kinked tails (100%) Meningocele (81%) Micrognathia (29%) Oligodactylia (10%) Syndactylia (4%) Anal atresia (3%)	Teramoto et al., 1978a
Syrian golden hamster	0, 90, 270, 810 mg/kg oral	11–18	6–13	14	None	None 90 mg/kg: No findings 270 mg/kg: Shortened and kinked tails (14%) 810 mg/kg: number of live foetuses significantly reduced Foetal weight significantly reduced malformations (63%) Cleft palate, kinked tail, oligodactylia, anal atresia, vertebral malformations, scoliosis	Teramoto et al., 1978a
Golden hamster	—	0, 60, 120, 300 or 600 mg/kg oral	7–14	—	None	60 mg/kg: No findings 120 mg/kg: Lobes of the lung reduced, fused ribs and strenebrae, hydroureter 300 mg/kg: lobes of the lung reduced, fused ribs and sternebrae, hydroureter	Lu et al., 1980

Table 7 (continued)

Animal species and breed	Number per group	Dose and route of administration	Day of pregnancy Treatment	Day of pregnancy Sacrificed	Maternal toxicity	Findings in foetuses	References
	—	0, 300 or 360 mg/kg oral	7–11		Resorptions	600 mg/kg: Cleft palate 300 and 600 mg/kg: Increased resorption rate, cleft palate, micrognathia, hydroureter, fused ribs, missing kidneys, exencephaly, meningocele, anophthalmia	
	—	0 or 360 mg/kg	11–12	—	Resorptions	360 mg/kg: Increased resorption rate, cleft palate, oligodactylia, micrognathia, hydroureter, fused ribs, missing kidneys	
Rat, Sprague-Dawley	113	0, 60, 120, 240 mg/kg	11	20	None	60 mg/kg: No findings 120 mg/kg: Myeloschisis (26%) 240 mg/kg: Myelomeningocele (44%)	Hung et al., 1986
			12	20	None	60 mg/kg: No findings 120 mg/kg: Exencephaly (43%) Meningoencephalocele (32%) 240 mg/kg: Exencephaly (83%) Meningoencephalocele (17.4%)	
			13	20	None	60 mg/kg: Exencephaly (54%) Meningoencephalocele (38%)	

Ethylene thiourea 123

14	20	None	120 mg/kg: Exencephaly (60%) Meningoencephalocele (40%) 240 mg/kg: Exencephaly (48%) Meningoencephalocele (41%) 60 mg/kg: Microencephaly (0%) Meningoencephalocele (51%) 120 mg/kg: Microencephaly (77%) Meningoencephalocele (23%) 240 mg/kg: Microencephaly (30%) Meningoencephalocele (70%)
15	20	None	60 mg/kg: Hydroencephaly (100%) 120 mg/kg: Hydroencephaly (100%) 240 mg/kg: Hydroencephaly (100%)
16	20	None	60 mg/kg: Hydroencephalocele (100%) 120 mg/kg: Hydroencephalocele (100%) 240 mg/kg: Hydroencephalocele (100%)
17, 18	20	None	60 mg/kg: Hydrocephalus (100%) 120 mg/kg: Hydrocephalus (100%) 240 mg/kg: Hydrocephalus (100%)

References

Allen, J.R., Van Miller, J.P., Seymour, J.L.
Absorption, tissue distribution and excretion of ^{14}C ethylenethio-urea by the Rhesus monkey and rat
Res. Comm. Chem. Pathol. Pharmacol., 20, 109–115 (1978)

Arnold, D.L., Krewski, D.R., Junkins, D.B., Mcbuire, P.F., Moodie, C.A., Munro, J.C.
Reversibility of ethylenethiourea-induced thyroid lesions
Toxicol. Appl. Pharmacol., 67, 264–273 (1983)

Autio, K., Wright, A., Pyysalo, H.
The effects of oxidation of the sulfur atom on the mutagenicity of ethylenethiourea
Mutat. Res., 106, 27–31 (1982)

Bruze, M., Fregert, S.
Allergic contact dermatitis from ethylenethiourea
Contact Dermatitis, 9, 208–212 (1983)

Chernoff, N., Kavlock, R.J., Rogers, E.H., Carver, B.D., Murray, S.
Perinatal toxicity of maneb, ethylene thiourea and ethylenebisisothiocyanate sulfide in rodents
J. Toxicol. Environ. Health, 5, 821–834 (1979)

Doull, J., Plzak, V., Brois, S.J.
A survey of compounds for radiation protection
School of aerospace medicine, USAF aerospace medical division (AFSC)
Brooks air force base, Texas, 1962
National Technical Information Service, AD-277 689

Du Pont
Chemicals for Elastomers
Bulletin 14 A vom 12. Oktober 1973

Freudenthal, R.I., Kerchner, G., Persing, R., Baron, R.L.
Dietary subacute toxicity of ethylene thiourea in the laboratory rat
J. Environ. pathol. Toxicol., 1, 147–161 (1977)

Gak, J.-C., Graillot, C., Truhaut, R.
Différence de sensibilité du hamster et du rat vis-à-vis des effets de l'administration à long terme de l'éthylène thiorée
Europ. J. Toxicol., 9, 303–312 (1976)

Graham, S.L., Hansen, W.H., Davis, K.J., Perry, C.H.
Effects of one-year administration of ethylenethiourea upon the thyroid of the rat
J. Agr. Food Chem., 21, 324–329 (1973)

Graham, S.L., Davis, K.J., Hansen, W.H., Graham, C.H.
Effects of prolonged ethylene thiourea ingestion on the thyroid of the rat
Cosmet. Toxicol, 13, 493–499 (1975)

Graham, S.L., Hansen, W.H.
Effects of short-term administration of ethylenethiourea upon thyroid function of the rat
Bull. Environ. Contam. Toxicol., 7, 19–25 (1972)

Hoechst AG
Mitteilung an die Berufsgenossenschaft der chemischen Industrie

Hui, Q.Y., Armstrong, C., Laver, G., Iverson, F.
Monooxygenase-mediated metabolism and binding of ethylenethiourea to mouse liver microsomal protein
Toxicol. Lett., 41, 231–237 (1988)

Hung, C.-F., Lin, K.-R., Lee, C.-S.
Experimental production of congenital malformation of the central nervous system in rat fetuses by single intragastric administration of ethylenethiourea
Proc. Natl. Sci. Counc. B. ROC, 10, 127–136 (1986)

IARC (International Agency for Research on Cancer)
Monographs on the Evaluation of the Carcinogenic Risk of Chemicals to Man: Some anti-thyroid and related substances, nitro-furans and industrial chemicals
Vol. 7, 45–52 (1974)
IARC Monographs, Supplement 4, 128–130 (1982)

Innes, J.R.M., Ulland, B.M., Valeio, M.G., Petrucelli, L., Fishbein, L., Hart, E.R., Pallotta, A.J., Bates, R.R., Falk, H.L., Gart, J.J., Klein, M., Mitchel, J., Peters, J.
Bioassay of pesticides and industrial chemicals for tumorigenicity in mice: A preliminary note
J. Nal. Cancer Inst., 42, 1101–1114 (1969)

Iverson, F., Khera, K.S., Hierlihy, S.L.
In vivo and in vitro metabolism of ethylenethiourea in the rat and in

the cat
Toxicol. Appl. Pharmacol., 52, 16–21 (1980)

Jordan, L.W., Neal, R.A.
Examination of the in vivo metabolism of Maneb and Zineb to ethylenethiourea (ETU) in mice
Bull. Environ. Contam. Toxicol., 22, 271 (1979)

Kato, Y., Odanaka, Y., Teramoto, S., Matano, O.
Metabolic fate of ethylenethiourea in pregnant rats
Bull. Environ. Contam. Toxicol., 16, 546–555 (1976)

Kavlok, R., Chernoff, N.
Postnatal toxicity of Maneb, Ebis and ETU in rats
Toxicol. Appl. Pharmacol., 45, 358 (1978)

Khera, K.S.
Ethylenethiourea: Teratogenicity study in rats and rabbits
Teratology, 7, 243–252 (1973)

Khera, K.S.
Neuronal degeneration caused by ethylenethiourea in neuronal monocell layers in vitro and fetal rat brain in vivo
Teratology, 33, 39 C (1986)

Khera, K.S., Iverson, F.
Effects of pretreatment with SKF-525A, N-methyl-2-thioimidazole, sodium phenobarbital or methylcholanthrene on ethylenethiourea-induced teratogenicity in rats
Teratology, 24, 131–137 (1981)

Khera, K.S., Iverson, F.
Toxicity of ethylenethiourea in pregnant cats
Teratology, 18, 311–314 (1978)

Khera, K.S., Tryphonas, L.
Ethylenethiourea-induced hydrocephalus: pre- and postnatal pathogenesis in offspring from rats given a single oral dose during pregnancy
Toxicol. Appl. Pharmacol., 42, 85–97 (1977)

Khera, K.S., Tryphonas, L.
Nerve cell degeneration and progeny survival following ethylenethiourea treatment during pregnancy in rats
Neurotoxicology, 6, 97 (1985)

Khera, K.S., Whalen, C., Iverson, F.
Effects of pretreatment with SKF-525A, N-methyl-2-thioimidazole, sodium phenobarbital or 3-methylcholanthrene on ethylenethiourea-induced teratogenicity in hamsters
J. Toxicol. Environ. Health, 11, 287–300 (1983)

Kobayashi, H., Kaneda, M., Teramoto, S.
Identification of 1-methylthiourea as the metabolite of ethylenethiourea in rats by HPLC
Toxicol. Lett., 12, 109–113 (1982)

Korhonen, A., Hemminki, K., Vainio, H.
Embryotoxicity of industrial chemicals on the chicken embryo: Thiourea derivatives
Acta Pharmacol. Toxicol., 51, 38–44 (1982)

Lehto, V.-P., Vistanen, J., Savolainen, K.
The effect of some dithiocarbamates, Disulfiram and 2,5-hexandione on the cytoskeleton of neuronal cells in vivo and in vitro
The Cytoskeleton
ed. by Thomas W. Clarkson, Polly R. Sager, Tore L.M. Syversen
Plenum Publishing Corporation, S. 143 (1984)

Lewerenz, H.J., Bleyl, D.W.R.
Postnatal effects of oral administration of ethylenethiourea in rats during late pregnancy
Arch. Toxicol., Suppl. 4, 292–295 (1980)

Lewerenz, H.J., Plass, R.
Contrasting effects of ethylenethiourea on hepatic mono oxygenase in rats and mice
Arch. Toxicol., 56, 92–95 (1984)

Lewerenz, H.J., Plass, R., Bleyl, D.W.R.
Untersuchungen zum Koergismus ausgewählter Pestizide
1. Mitt. Akute orale Toxizität bei kombinierter Applikation
Die Nahrung, 24, 463–469 (1980)

Lu, M.H., Staples, R.E.
Teratogenicity of ethylenethiourea and thyroid function in the rat
Teratology, 17, 171–178 (1978)

Lu, M.H., Su, C.T., Chang, F.Y.H., Hoopes, I.H.
Teratogenic potential of ethylenethiourea (ETU) in the hamster
Teratology, 21, 54 A (1980)

Matsushita, T., Arimatsu, Y., Nomura, Sh.
Experimental study on contact dermatitis caused by dithiocarbamates Maneb, Mancozeb, Zineb and their related compounds
Int. Arch. Occup. Environ. Health, 37, 169–178 (1976)

Mollet, P.
Toxicity and mutagenicity of ethylenethiourea (ETU) in Drosophila
Mutat. Res., 29, 254 (1975)

Newsome, W.H.
The excretion of ethylenethiourea by rat and Guinea pig
Bull. Environ. Contam. Toxicol., 11, 174–176 (1974)

NTP (National Toxicology Program)
Annual Plan for Fiscal Year 1983
NTP-82-119, January 1983

NTP (National Toxicology Program)
Annual Plan for Fiscal Year 1984
NTP-84-023, Febr.1984

NTP (National Toxicology Program)
Annual Plan for Fiscal Year 1985
NTP-85-055, March 1985

Plasterer, M.R., Bradshaw, W.S., Booth, G.M., Carter, M.W., Schuler, R.L., Hardin, B.D.
Developmental toxicity of nine selected compounds following prenatal exposure in the mouse: Naphthalene, p-nitrophenol, sodium selenite, dimethylphthalate, ethylenethiourea and four glycol ether derivatives
J. Toxicol. Environ. Health, 15, 25–38 (1985)

Römpp's Lexikon der Chemie (O.A. Müller),
7. Auflage, S. 1568
Frankhsche Verlagshandlung, Stuttgart (1977)

Ruddik, J.A., Khera, K.S.
Pattern of anomalies following single oral dosis of ethylenethio-urea to pregnant rats
Teratology, 12, 277–282 (1975)

Ruddik, J.A., Newsome, W.H., Iverson, F.
A metabolic comparison of ethylenethiourea in the rat and mouse
Teratology, 13, 34 A (1976)

Ruddik, J.A., Newsome, W.H., Iverson, F.
A comparison of the distribution, metabolism and excretion of ethylenethiourea in the pregnant mouse and rat
Teratology, 16, 159–162 (1977)

Ruddik, J.A., Williams, D.T., Hierlihy, L., Khera, K.S.
(^{14}C) Ethylenethiourea: Distribution, excretion and metabolism in pregnant rats
Teratology,13, 35–40 (1976)

Savolainen, K., Pyysalo, H.
Identification of the main metabolite of ethylenethiourea in mice
J. Agric. Food Chem., 27, 1177–1181 (1979)

Schüpbach, M., Hummler, H.
A comparative study on the mutagenicity of ethylenethiourea in bacterial and mammalian test systems
Mutat. Res., 56, 111–120 (1977)

Seiler, J.P.
Ethylenethiourea (ETU) a carcinogenic and mutagenic metabolite of ethylenebisdithiocarbamate
Mutat. Res., 26, 189–191 (1974)

Seiler, J.P.
Inhibition of testicular DNA synthesis by chemical mutagenes and carcinogens. Preliminary results in the validation of novel short term test
Mutat. Res., 46, 305–310 (1977)

de Serres, F.J., Ashby, J. (eds.)
Progress in Mutation Research, Vol. 1
Evaluation of Short-Term Tests for Carcinogens
Report of the International Collaborative Program, New York, Elsevier/North-Holland Biomedical Press
pp. 83, 129, 175, 183, 193, 199, 210, 351, 385, 396, 414, 424, 434, 457, 468, 481, 491, 502, 517, 528, 533, 538, 551, 560, 570, 580, 594, 626, 638, 651, 673, 686, 698, 712 (1981)

Smith, D.
Ethylene thiourea: a study of possible teratogenicity and thyroid carcinogenicity
J. Soc. Occup. Med., 26, 92–94 (1976)

Smith, D.
Ethylene thiourea: thyroid function in two groups of exposed workers
Br. J. Ind. Med., 41, 362–366 (1984)

Stula, E.F., Krauss, W.C.
Embryotoxicity on rats and rabbits from cutaneous application of amide-type solvents and substituted ureas
Toxicol. Appl. Pharmacol., 41, 35–55 (1977)

Teramoto, S., Kaneda, M., Aoyama, H., Shirasu, Y.
Correlation between the molecular structure of N-alkylureas and N-alkylthioureas and their teratogenic properties
Teratology, 23, 335–342 (1981)

Teramoto, S., Moriya, M., Kato, K., Tezuka, H., Nakamura, S., Shingu, A., Shirasu, Y.
Mutagenicity testing of ethylenethiourea
Mutat. Res., 56, 121–129 (1977)

Teramoto, S., Harada, T., Kaneda, M.
Effects of ethylenethiourea upon developing brain and limb buds in the rat
Cong. Anom., 18, 241–249 (1978b)

Teramoto, S., Shingu, A., Kaneda, M., Saito, R.
Teratogenicity studies with ethylenethiourea in rats, mice and hamsters
Cong. Anom., 18, 11–17 (1978a)

Ugazio, G., Brossa, O., Grignolo, F.
Hepato- and neurotoxicity by ethylenethiourea
Res. Commun. Chem. Pathol. Pharmacol., 48, 401–414 (1985)

Ulland, B.M., Weisburger, J.H., Weisburger, E.K., Rice, J.M., Cypher, R.

Brief Communication: Thyroid cancer in rats from ethylene thiourea intake
J. Natl. Cancer Inst., 49, 583–584 (1972)

Weisburger, E.K., Ulland, B.M., Nam, J., Gart, J.J., Weisburger, J.H.
Carcinogenicity tests of certain environmental and industrial chemicals
J. Natl. Cancer Inst., 67, 75–88 (1981)

Woodruff, R.C., Mason, I.M., Valencia, R., Zimmering, S.
Chemical mutagenesis testing in Drosophila
V. Results of 53 coded compounds tested for the National Toxicology Program
Environ. Mutagen., 7, 677–702 (1985)

Gamma-Butyrolactone

1. Summary and assessment

Gamma-Butyrolactone (GBL) is rapidly metabolized via GHB and 66%–80% is eliminated in the breath as carbon dioxide within 6–18 hours. It has a week narcotic effect; this is due to a fast metabolic conversion of GBL to Gamma-hydroxybutyric acid (GHB), which has an effect on the central nervous system.

GBL is of relatively low acute toxicity (LD_{50} in the rat orally 1.5–1.8 g/kg). No toxic effects, apart from some loss of weight, have been observed in animal experiments using repeated administration of high doses (3 g/kg in drinking water daily for 4 weeks).

GBL does not irritate the skin; it has, however, a definite irritative effect on the conjunctiva of the eye. The reported damage to the cornea and iris are reversible. The substance can be absorbed percutaneously. No indications of a sensitizing effect can be deduced from the available reports.

The literature offers no evidence of GBL having any chronic toxic effects.

The large numbers of in vitro and in vivo experiments carried out with GBL have yielded no indication of a genotoxic effect of this substance.

From long-term studies on mice there is no evidence of a carcinogenic effect of GBL following oral, subcutaneous or dermal administration. In rats, the experiments involving oral or subcutaneous administration cannot be assessed because of the small numbers of animals involved and the inadequate information regarding control groups. A 2-year oral study in rats and mice is presently at the stage of histopathological evaluation.

In an experimental study on 200 human volunteers, no primary irritative effect from undiluted GBL was found, and there were no indications of a positive sensitizing action on the skin.

2. Name of substance

2.1 Usual name   Gamma-Butyrolactone
2.2 IUPAC-name   Dihydro-2(3H)-furanone

Gamma-Butyrolactone

2.3 CAS-No. 96-48-0

3. Synonyms

Common and trade names
GBL
gamma-BL
BLO
1,4-Butanolide
Butyric acid 4-hydroxy-, gamma-lactone
Butyric acid lactone
4-Butyrolactone
Butyryl lactone
4-Hydroxybutanoic acid lactone
4-Hydroxybutyric acid, gamma-lactone
4-Hydroxybutanoic acid, gamma-lactone
gamma-Hydroxybutyric acid cyclic ester
4-Hydroxybutyric acid lactone
gamma-Hydroxybutyric acid lactone
NCI-C55878

4. Structural and molecular formulae

4.1 Structural formula

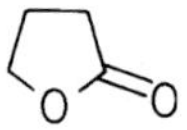

4.2 Molecular formula $C_4H_6O_2$

5. Physical and chemical properties

5.1 Molecular mass 86.1 g/mol
5.2 Melting point −44° C (solidification point)
5.3 Boiling point 204–206° C
5.4 Density 1.13 g/cm^3 (at 20° C)

5.5 Vapour pressure	1.5 hPa (at 20° C)
	10.7 hPa (at 60° C)
	52.5 hPa (at 100° C)
5.6 Solubility in water	miscible with water in any amount
5.7 Solubility in organic solvents	miscible with ethanol, ether, acetone, benzene in any amount
5.8 Solubility in fat	limited solubility in soya oil
5.9 pH-value	no information available
5.10 Conversion factors	1 ppm $\hat{=}$ 3.57 mg/m^3
	1 mg/m^3 $\hat{=}$ 0.28 ppm
	(BASF, 1981; IARC, 1976; Kronevi, 1988; Ullmann, 1975).

6. Uses

Intermediate in the manufacture of 2-pyrrolidone, N-methylpyrrolidone and N-vinylpyrrolidone; solvent for polymers; intermediate in the manufacture of the herbicide 4-(2,4-dichlorophenoxy)butyric acid (IARC, 1976; Ullmann, 1975).

7. Results of experiments

7.1 Toxicokinetics and metabolism

Gamma-Butyrolactone (GBL) is quickly metabolized in the rat to Gamma-hydroxybutyric acid (GHB). The half-life in rat blood in vitro for this conversion was <1 minute. The metabolism was accomplished by the enzyme lactonase, which could be detected mainly in the serum and in the liver (Roth and Giarman, 1965, 1966).

When equimolar doses of GBL (500 mg/kg) or GHB-Na salt (732 mg/kg) were injected i.v. into male Sprague-Dawley rats, the total blood level immediately after administration of GBL was approx 50% lower than after administration of GHB. The blood concentration of GBL fell from 13 µmol/ml (1-minute value) to 6.5 µmol/ml after 45 minutes. Within 3 hours the blood levels of GBL and GHB approached zero, the decline of GBL being slower than that of GHB (Roth and Giarman, 1965).

^{14}C-Carboxyl-labelled GBL and GHB were quickly metabolized to ^{14}CO$_2$ in vivo. After a single i.v. dose of 0.365 µM (2 µc) of ^{14}C-GHB in rats, the first traces of $^{14}CO_2$ could be detected in the respiratory air after less than 4 minutes. A maximum was reached after 15 minutes. 60% of the total ^{14}C radioactivity was eliminated as $^{14}CO_2$ in the respiratory air within 2.5 hours. After injection of the same dose of ^{14}C-GBL (2 μc) the maximum ^{14}C concentration in the respiratory air occurred after just 20 minutes, with expiration of $^{14}CO_2$ taking place similarly as for ^{14}C-GHB. The time difference between occurrence of the $^{14}CO_2$ maxima after administration of GHB and GBL was presumed by the authors to result from the time required for the metabolism of GBL to GHB by the organism (Roth and Giarman, 1966).

To clarify the question of which organ is involved in the breakdown of GHB to CO_2, sections of brain and liver were incubated with variable doses of ^{14}C-GHB. Both organs metabolized GHB, but the liver sections were twice as active as the brain sections (Roth and Giarman, 1966).

When (1-^{14}C)- or (4-^{14}C)-hydroxybutyrate was administered i.p. to rats at a dose of 40 mg/kg, 66% of the radioactivity was eliminated as $^{14}CO_2$ in 6 hours, and another 10–20% in 18 hours. No differences were seen in CO_2 exhalation after administration of (1-^{14}C)- and (4-^{14}C)- hydroxybutyrate (Walkenstein et al., 1964).

When 6.34 mM GBL/kg was administered percutaneously to male Sprague-Dawley rats, GHB could be detected in the blood at concentrations of up to 200 μg/ml. The time course of percutaneous absorption depended on previous treatment of the rat skin: if the skin had only been shaved, the blood level of GHB increased slowly, reached a maximum after 2 hours and then decreased again. If, however, the skin was treated with a commercial depilatory after shaving, the blood level of GHB reached its maximum approx. 10 minutes after application of GBL, and then decreased slowly. It was deduced from the results of the investigations that at least 10% of the applied dose penetrated the skin (Fung et al., 1979).

7.2 Acute and subacute toxicity

The values for acute toxicity (LD_{50}) are summarized in the following table 1:

Table 1. Acute toxicity (LD_{50}) of Gamma-Butyrolactone in different species

Species	Route of administration	LD_{50}	References
Rat	oral	ca 1.58 g/kg	BASF, 1981
Rat	oral	1.8 g/kg	Kvasov, 1974
Mouse	oral	0.8–1.6 g/kg	Fassett, 1963
Rat	oral	1.5 ml/kg	Ind. Toxicol. Lab., 1952
Mouse	oral	1.26 g/kg	Hampel et al., 1968
Guinea pig	oral	1.5 ml/kg	Ind. Toxicol. Lab., 1952
Guinea pig	oral	0.5–0.7 g/kg	Freifeld and Horst, 1967
Rat/mouse	i.p.	0.2–0.4 g/kg	Fassett, 1963
Mouse	i.p.	0.88 g/kg	Hampel et al., 1968
Guinea pig	dermal	ca 5.6 g/kg	Fassett, 1963
Rat	Inhalation risk test, 8 hours at 20° C, saturated atmosphere	No deaths, no symptoms of intoxication	BASF, 1981
Mouse	i.v.	880 mg/kg	Hampel et al., 1968

Apart from a slight narcotic effect, no characteristic symptoms of intoxication were reported. At doses of 100–200 mg/kg i.p., GBL had a slight narcotic effect in rats. After an initial reduction in locomotor activity, marked hyperactivity and restlessness were observed (Davies, 1978).

The narcotic effect in the rat lasted for approx 90 minutes at 200 mg/kg i.p., for 3 hours at 400 mg/kg i.p., and for 5–8 hours at 700–800 mg/kg i.p. (Borbély and Huston, 1972).

Application of approx 3 g/kg per day via the drinking water (2% GBL solution) over a period of 4 weeks caused only a slight but significant reduction in weight gain ($p < 0.05$) in male Sprague-

Dawley rats. The tolerance of this relatively high daily dose, compared with the acute oral LD_{50}, was explained by the rapid metabolism of the substance in the animal's body. When the test and control animals were injected i.p. with single GBL doses of 350 and 750 mg/kg at the end of the test period, the lower dose resulted in a 58% reduction in the duration of narcosis in the rats treated over a period of 4 weeks with approx 3 g/kg/day. In comparison with the control rats, with administration of 750 g/kg i.p. the corresponding reduction was 11% (Nowycky and Roth, 1979).

7.3 Skin and mucous membrane irritation

Undiluted GBL did not cause skin irritation even after a 20-hour application (semi-occlusive) to the very sensitive skin of the back of white rabbits (BASF, 1981).

A slight irritant effect on the skin of the guinea pig has been reported (without further details; Lazarev, 1969).

On the other hand, after instillation of 50 µl in the conjunctival sac of the rabbit eye, there was severe irritation of the conjunctivae (BASF, 1981).

In another eye irritation test, lesions were observed in the cornea, iris and conjunctivae after instillation of undiluted GBL in the conjunctival sac. The changes in the cornea and iris had largely receded within 7 days (Food and Drug Research Lab., 1977).

7.4 Sensitization

In guinea pigs, no indications of skin-sensitizing action were seen in tests which were not described in more detail (Fassett, 1963).

7.5 Subchronic-chronic toxicity

Rats (FDRL breed, 20 males and 20 females/group) and beagles (3 males and 3 females/group) received 0.2, 0.4 and 0.8% GBL in the feed over a period of 90 days. Corresponding controls were also set up. The results showed that GBL at the concentrations tested did not produce any symptoms of intoxication or pathological effects. The parameters investigated were: behaviour, appearance, survival, growth, feed intake, haematological and clinical-chemical investigations, analysis of urine and histopathological examinations (Food and Drug Research Lab., 1979).

7.6 Genotoxicity

GBL did not exhibit any alkylation reaction with guanosine, RNA, DNA or 4-(p-nitrobenzyl)-pyridine (Hemminki, 1981).

When GBL was tested in the SCE (Sister Chromatid Exchange) test in CHO (Chinese Hamster Ovary) cells, negative results were found without metabolic activation at doses of up to 1500 µg/ml, but with metabolic activation positive results were found at doses of 3000–5000 µg/ml (NTP, 1986).

Chromosome aberrations in CHO cells were not seen at doses of up to 5000 µg/ml without metabolic activation, but with metabolic activation they occurred at 2600 µg/ml and above (NTP, 1986).

Within the scope of the "Evaluation of Short Term Tests for Carcinogens, Report of the International Collaborative Program" (Progress in Mutation Research, Vol. 1, 1981), GBL was also tested in repair tests, in mutation assays with bacteria and yeasts, and in tests with animal cells, both in vitro and in vivo.

The following tests were reported:

Table 2. Repair test

Test	Result	Author
Bacillus subtilis Rec assay with S9	Positive	Kada, 1981
E. coli RecA$^-$/PolA$^-$	Negative	Green, 1981
E. coli RecA$^-$	Negative	Ishinotsubo et al., 1981a
E. coli RecA$^-$/PolA$^-$	Negative	Tweats, D., 1981
E. coli PolA without S9	Uncertain	Rosenkranz, et al., 1981.
Lambda induction (lysis) in E. coli	Negative	Thomson, 1981

Apart from the Rec assay with Bacillus subtilis, the findings of various investigators in the repair test with E. coli were negative.

Table 3. Point mutations in bacteria

Test	Result	Author
Salmonella/microsome test ± S9	Negative	Baker and Bouin, 1981
Salmonella/microsome test ± S9	Negative	Broods and Dean, 1981
Salmonella/microsome test ± S9	Negative	Martire, G. et al., 1981
Salmonella/microsome test ± S9	Negative	Garner et al., 1981
Salmonella/microsome test ± S9	Negative	MacDonald, 1981
Salmonella/microsome test ± S9	Negative	Ishinotsubo et al., 1981b
Salmonella/microsome test ± S9	Negative	Nagao and Takahashi, 1981
Salmonella/microsome test ± S9	Negative	Richold and Jones, 1981
Salmonella/microsome test ± S9	Negative	Rowland and Severn, 1981
Salmonella/microsome test ± S9	Negative	Simmon and Shepherd, 1981
Salmonella/microsome test ± S9	Negative	Trueman, 1981
Salmonella/microsome test ± S9	Negative	Venitt and Crofton-Sleigh, 1981
Fluctuation test with TA 98 and TA 100 ± S9 and hepatocytes	Negative	Hubbard, 1981
Salmonella typhimurium strain TM 677 (8-azaguanine resistance) ± S9	Negative	Skopek et al., 1981
"Mitotic" fluctuation test with E. coli WP 2 uvrA, Salmonella typhimurium TA 98, TA 1535, TA 1537 ± S9	Negative	Gatehouse, 1981
Reversion mutation test with the E. coli WP 2 system (pre-incubation) ± S9	Negative	Matsushima, 1981

No genotoxic effect was detected in the various in vitro investigations on bacteria with and without S9-mix.

Table 4. Tests on yeasts

Test	Result	Author
Sacch. cerevisiae T1, T2 (mitotic crossing over)	Negative	Kassinova et al., 1981
Sacch. cerevisiae T3, T4 (repair-deficient strains) with and without S9	Negative	Kassinova et al., 1981
Sacch. cerevisiae D7 (induction of mitotic gene conversions) with S9	Negative	Zimmermann and Scheel, 1981
Sacch. cerevisiae D4 (induction of mitotic gene conversions) with S9	Negative	Jagannath et al., 1981
Sacch. cerevisiae XI-185-14C (induction of reversions) ± S9	With S9 uncertain Without S9 negative	Mehta and von Borstel, 1981
Sacch. pombe P1 (induction of forward mutations)	Negative	Loprieno, 1981
Sacch. cerevisiae D6 (induction of mitotic aneuploidy)	Negative	Parry and Sharp, 1981
Sacch. cerevisiae JD1 (induction of mitotic gene conversions)	Negative	Sharp and Parry, 1981a
Sacch. cerevisiae rad. (test for inhibition of growth in wild types in comparison with rad (repair-deficient types)	Negative	Sharp and Parry, 1981b

Testing of the genotoxic action of GBL on various yeasts did not show any indication of a mutagenic effect.

Table 5. In vitro tests with mammalian cells

Test	Result	Author
UDS HeLa cells	Negative	Martin and McDermid, 1981
Chromosome aberrations in RL1 cells	Negative	Dean, 1981
V 79 cells (HGPRT)	Negative	Knaap et al., 1981
Human fibroblasts Diphteria toxin resistance (HFDipr)	Negative	Gupta and Goldstein, 1981
BHK-21 cell transformation	Positive	Styles, 1981
SCE test in CHO cells only with S9-mix at high doses (3000–5000 µg/ml)	Positive	NTP, 1986
Chromosome aberrations in CHO cells only with S9-mix at high doses (starting from 2600 µg/ml)	Positive	NTP, 1986

GBL was negtative in the UDS test, in the chromosome aberration test and in the HGPRT test (V 79 cells), on the other hand the cell transformation test, the SCE test and the chromosome aberration test were positive at high doses with S9-mix.

Table 6. In vivo tests

Test	Result	Author
Drosophila melanogaster	Negative	Vogel et al., 1981
Micronucleus test on B6C3F1 mice	Negative	Salamone et al., 1981
Micronucleus test on CD1 mice	Negative	Tsuchimoto and Matter, 1981
Indication of abnormal spermatozoa	Negative	Topham, 1981

These in vivo test systems did not show any indications of a genotoxic effect for GBL.

7.7 Carcinogenicity

A group of 60 four-week old C3H mice (30 male and 30 female) received, throughout their life, feed to which 1 g GBL/kg had been added. Another group of 36 XVII/G mice (male and female) received, again throughout their life, a dose of 2 mg /GBL/0.1 ml water/animal by stomach tube twice weekly. In the female C3H mice, mammary tumours developed in 43 out of 61 untreated animals (70.5%), and in 19 out of 30 treated animals (63.3%). The mean latency period for the development of tumours was 327 ± 59 days in the controls, and 315 ± 63 days in the treated females. In 54 males in the control group, hepatomas were observed in 9% of the animals (6 out of 54), whereas the frequency of these tumours in the treated animals was 16.6% (5 out of 30). The difference in the two frequencies was not statistically significant ($p > 0.05$). In the 36 treated XVII/G mice the incidence of lung tumours was 55% (20 out of 36 mice; mean survival time: 571 days), compared with 61% (27 out of 44) of the untreated control animals (mean survival time: 595 days; Rudali et al., 1976).

95 male NMRI mice (8 weeks old) received 750 mg GBL/kg orally once per week for 18 months, and a solvent control (95 male mice) was also conducted. With regard to tumour frequency – mainly lymphomas and lung adenomas were found – there was no statistically significant difference between the treated group and the control group. In the mice treated with GBL, histological investigation did not reveal any pathological changes that differed from the controls (Holmberg et al., 1983).

34 newborn XVII/G mice received s.c. injections of 1 µg GBL on the 1st, 4th and 8th day of life. 53% developed lung tumours (mean life: 590 days) whereas out of 44 untreated controls, 61% had lung tumours (mean life: 595 days; Rudali et al., 1976).

16 female Swiss-Webster mice received 0.005 mg GBL and 0.1 ml tricaprylin subcutaneously 3 times weekly over a period of 4 weeks (12 injections). 11 mice survived 18 months. No local tumours were found (Swern et al., 1970).

30 male Swiss-Millerton (ICR/Ha) mice, approx. 8 weeks old, were treated on the shaved skin of the back 3 times weekly for their whole life with 0.1 ml of a 10% GBL solution in benzene. Two of these animals (approx. 6.9%) developed skin tumours, one of them a carcinoma. The mean life of the animals of this group was 292 days. Of 150 benzene-treated control mice, 11 developed skin tumours (approx. 8%), and one of them had a skin carcinoma. The mean life

in these 4 control groups was 262–412 days (van Duuren et al., 1963).

30 male Swiss-Millerton (ICR/Ha) mice were painted 3 times weekly for their whole life on the shaved skin of the back with 0.1 ml of a 10% solution of GBL in acetone. The mice had a mean survival time of 495 days, and there were no skin tumours and no local irritation (van Duuren et al., 1965).

30 XVII/G mice aged 4 weeks received one drop of a 1% GBL solution in acetone on the skin of the nape, twice weekly for their whole life. A control group of 17 mice was treated with acetone alone. Lung tumours occurred in both groups: 70% in the GBL group, 52,9% in the acetone group. The survival times of the GBL group averaged 601 days, and of the acetone group 499 days. Skin tumours were not observed (Rudali et al., 1976).

30 ICR/Ha mice (male and female) were treated for 42 weeks, 3 times weekly with 10 mg GBL applied to the skin (no precise details). Papillomas developed in 2 animals. The test result was assessed as negative (no further details; Searle, 1976).

Newly-weaned male white rats (no information on number and breed) received 4–6 doses of 100–400 mg/kg (total doses 450–1700 mg/kg) of GBL by stomach tube for 7 months. Out of 7 rats which survived for 18–28 months from the start of treatment, 5 developed tumours (2 hypophysis tumours, 2 epithelial cell carcinomas of the jaw and 1 interstitial tumour of the testes). Controls were not conducted. According to the authors, tumours in the jaw and testes seldom occur in control rats (Schoental, 1968). Because of the small number of animals, and in view of the fact that GBL was obtained by distillation from a mixture of GBL and 4,4′-diaminodiphenylmethane, the purity only having been documented by the boiling point, the results cannot be evaluated.

5 male Wistar rats (8 weeks old) received 2 mg as a solution in oil injected subcutaneously twice weekly over a period of 61 weeks. The animals were observed for a total of 100 weeks. No tumours were seen (Dickens and Jones, 1961).

A 2-year study involving oral administration to rats and mice is currently undergoing histopathological evaluation (NTP, 1988).

7.8 Reproductive toxicity

Female Sprague-Dawley rats received GBL orally in soya oil at doses of 10, 50, 125, 250, 500 and 1000 mg/kg/day from the 6th to the 15th day of pregnancy. Starting from 500 mg/kg, GBL had unsufficient solubility in soya oil. Body weights, feed intake and water intake were not altered in the mother animals, but the weight of the placenta was reduced in all test groups. There were no differences from the control groups in the corpora lutea, total implantations, ratio of dead to live foetuses, resorptions and pre- and post-implantation losses. The average foetal weight was significant increased at doses of 50, 125 and 250 mg/kg. There were no visceral or skeletal malformations caused by the substance (Kronevi, 1984; Kronevi et al., 1988).

7.9 Effects on the immune system
No information available.

7.10 Other effects

The purpose of a study was to test whether GBL or GHB is responsible for the narcotic effect that occurs after administration of GBL. It was found that there is a correlation between the occurrence of the narcotic effect and the concentration of GHB, but not of GBL, in the brain (Guidotti and Ballotti, 1970).

8. Experience in humans

200 male and female volunteers underwent a repeat patch test in order to test for any skin-sensitizing effect. A test pad soaked with undiluted GBL remained on the skin for 5 days. After this exposure the skin was examined for reddening and oedema. 3 weeks later the test was repeated with a contact time of 48 hours. No skin reactions occurred, either after the first or subsequent applications (Ind. Toxicol. Lab., 1950).

References

Baker, R.S.V., Bouin, A.M.
Study of 42 coded compounds with the Salmonella/mammalian microsome assay
Prog. Mutat. Res., 1, 250–260 (1981)

BASF AG
Data sheet "Butyrolacton" (1981)

Borbély, A.A., Huston J.P.
Gamma-Butyrolactone: An anesthetic with hyperthermic action in the rat
Experientia, 28, 1455 (1972)

Broods, R.M., Dean, B.J.
Mutagenic activity of 42 coded compounds in the Salmonella/microsome assay with pre-incubation
Prog. Mutat. Res., 1, 261–270 (1981)

Davies, J.A.
The effects of Gamma-Butyrolactone on locomotor activity in the rat
Psychopharmacol., 60, 67–72 (1978)

Dean, B.J.
Activity of 27 coded compounds in the RL1 chromosome assay
Prog. Mutat. Res., 1, 570–579 (1981)

Dickens, F., Jones, H.E.H.
Carcinogenic activity of a series of reactive lactones and related substances
Brit. J. Cancer, 15, 85–100 (1961)

Fassett, D.W.
Gamma-Butyrolactone (1963)
In: IARC Monographs on the Evaluation of Carcinogenic Risk of Chemicals to Man, 11, 231–239 (1976)

Food and Drug Research Laboratories
Submission of data by GAF Chemicals Corporation (1970)
Subacute feeding studies with butyrolactone in rats and dogs

Food and Drug Research Laboratories
Submission of data by GAF Chemicals Corporation (1977)
Rabbit eye irritation study: BLO

Freifeld, M., Hort, E.V.
1,4-Butylene glycol and gamma-butyrolactone (1967)
In: Kirk, R.E. and Othmer, D.F., eds.
Encyclopedia of Chemical Technology, 2nd ed., Vol. 10, pp. 667–676 (1967)
J. Wiley and Sons, New York

Fung, H.-L., Lettieri, J.T., Bochner, R.
Percutaneous butyrolactone absorption in rats
J. Pharm. Sci., 68, 1198–1200 (1979)

Garner, R.C., Welch, A., Pickering, C.
Mutagenic activity of 42 coded compounds in the Salmonella/microsome assay
Prog. Mutat. Res., 1, 280–284 (1981)

Gatehouse, D.
Mutagenic activity of 42 coded compounds in the "Microtiter" fluctuation test
Prog. Mutat. Res., 1, 376–386 (1981)

Green, M.H.L.
A differential killing test using an improved repair-deficient strain of Escherichia coli
Prog. Mutat. Res., 1, 183–194 (1981)

Guidotti, A., Ballotti, P.L.
Relationship between pharmacological effects and blood and brain levels of gamma-butyrolactone and gamma-hydroxybutyrate
Biochem. Pharmacol., 19, 883–894 (1970)

Gupta, R.S., Goldstein, S.
Mutagen testing in the human fibroblast diphtheria toxin resistance (HFDipr) system
Prog. Mutat. Res., 1, 614–625 (1981)

Hampel, H., Hapke, H.-J.
Arch. Int. Pharmacodyn. Ther., 171 (2), 306 (1968)
In: National Library of Medicine's Hazardous Substances Data Bank, May 28, 1987 on-line search

Hemminki, K.
Reactions of beta-propiolactone, beta-butyrolactone and gamma-butyrolactone with nucleic acids
Chem. Biol. Interactions, 34, 323–331 (1981)

Holmberg, B., Kronevi, T., Ackevi, S., Ekner, A.
The testing of carcinogenic activity in diphenylamine and gamma-butyrolactone by peroral administration in male mice
Arbete och Hälsa, 34, 1–35 (1983)

Hubbard, S.A.
Fluctuation test with S9 and hepatocyte activation
Prog. Mutat. Res., 1, 361–370 (1981)

IARC (International Agency for Research on Cancer
Monographs on the Evaluation of the Carcinogenic Risk of Chemicals to Man: Gamma-Butyrolactone
Vol. 11, 231–239 (1976)

Industrial Toxicology Laboratories
Submission of data by GAF Chemicals Corporation (1950)
Human repeated insult patch test: gamma-butyrolactone

Industrial Toxicology Laboratories
Submission of data by GAF Chemicals Corporation (1950)
Acute oral toxicity in the rat and guinea pig: gamma-butyrolactone

Ishinotsubo, D., Mower, H., Mandel, M.
Testing of a series of paired compounds (carcinogen and noncarcinogenic structural analog) by DNA repair-deficient E. coli strain
Prog. Mutat. Res., 1, 195–198 (1981a)

Ishinotsubo, D., Mower, H., Mandel, M.
Mutagen testing of a series of paired compounds with the Ames Salmonella testing system
Prog. Mutat. Res., 1, 298–301(1981b)

Jagannath, D.R., Vultaggio, D.M., Brusick, D.J.
Genetic activity of 42 coded compounds in the mitotic gene conversion assay using Saccharomyces cerevisiae Strain b4
Prog. Mutat. Res., 1, 456–467 (1981)

Kada, T.
The DNA-damaging activity of 42 coded compounds in the Rec Assay
Prog. Mutat. Res., 1, 175–182 (1981)

Kassinova, C.V., Kovaltsova, S.V., Marfin, S.V., Zakharov, I.A.
Activity of 40 coded compounds in differential inhibition and mitotic crossing-over assays in yeast
Prog. Mutat. Res., 1, 434–455 (1981)

Knaap, A.G.A.C., Goze, C., Simons, J.W.I.M.
Mutagenic activity of seven coded samples in V 79 chinese hamster cells
Prog. Mutat. Res., 1, 608–613 (1981)

Kronevi, T.
Presentation to the International Conference of Organic Solvent Toxicity, October 15–17, 1984, Stockholm, Sweden (1984)

Kronevi, T., Holmberg, B., Arvidsson, S.
Teratogenicity test of gamma-butyrolactone in Sprague-Dawley rats
Pharmacol. Toxicol., 62, 57–58 (1988)

Kvasov, A.R.
Toxicological characteristics of gamma-butyrolactone and 2-pyrrolidone as industrial poisons
Sb. Nauchn. Tr. Rostov. na-Donu Gos. Med. Inst., 17, 84–87 (1974)

Lazarev, N.V.
Harmful Substances in Industry (1969)
In: Kvasov, A.R.
Sb. Nauchn., Tr. Rostov. na-Donu Gos.Med. Inst., 17, 84–87 (1974)

Loprieno, N.
Screening of coded carcinogenic/noncarcinogenic chemicals by a forward-mutation system with the yeast Schizosaccaromyces pombe
Prog. Mutat. Res., 1, 424–433 (1981)

McDonald, D.J.
Salmonella/microsome test on 42 coded chemicals
Prog. Mutat. Res., 1, 285–297 (1981)

Martin, C.N., McDermid, A.C.
Testing of 42 coded compounds for their ability to induce unscheduled DNA repair synthesis in HeLa cells
Prog. Mutat. Res., 1, 533–537 (1981)

Martire, G., Vricella, G., Perfumo, A.M., DeLorenzo, F.
Evaluation of the mutagenic activity of coded compounds in the Salmonella test
Prog. Mutat. Res., 1, 271–279 (1981)

Matsushima, T., Takamoto, Y., Shirai, A., Sawamura, M., Sugimura, T.
Reverse mutation test on 42 coded compounds with the E. coli WP2 system
Prog. Mutat. Res., 1, 387–395 (1981)

Metha, R.D., von Borstel, R.C.
Mutagenic activity of 42 encoded compounds in the haploid yeast

reversion assay, strain XV 185–149
Prog. Mutat. Res., 1, 414–423 (1981)

Nagao, M., Takahashi, Y.
Mutagenic activity of 42 coded compounds in the Salmonella/microsome assay
Prog. Mutat. Res., 1, 302–313 (1981)

NTP (National Toxicology Program)
Fiscal Year 1986 Annual Plan
U.S. Department of Health and Human Services (1986)

NTP (National Toxicology Program)
Chemical Status Report dated 12.10.1988

Nowycky, M.-C., Roth, R.H.
Chronic gamma-butyrolactone (GBL) treatment: A potential model of dopamine hypoactivity
Naunyn-Schmiedebergs Arch. Pharmacol., 309, 247–254 (1979)

Parry, J.M., Sharp, D.C.
Induction of mitotic aneuploidy in the yeast strain D6 by 42 coded compounds
Prog. Mutat. Res., 1, 468–480 (1981)

Richold, M., Jones, E.
Mutagenic activity of 42 coded compounds in the Salmonella/microsome assay
Prog. Mutat. Res., 1, 314–322 (1981)

Rosenkranz, H.S., Hyman, J., Leifer, Z.
DNA Polymerase deficient assay
Prog. Mutat. Res., 1, 210–218 (1981)

Roth, R.H., Giarman, N.J.
Preliminary report on the metabolism of gamma-butyrolactone and gamma-hydroxybutyric acid
Biochem. Pharmacol., 14, 177–178 (1965)

Roth, R.H., Giarman, N.J.
Gamma-butyrolactone and gamma-hydroxybutyric acid distribution and metabolism
Biochem. Pharmacol., 15, 1333–1348 (1966)

Rowland, J., Severn, B.
Mutagenicity of carcinogens and noncarcinogens in the Salmonel-

la/microsome test
Prog. Mutat. Res., 1, 323–332 (1981).

Rudali, G., Apiou, F., Boyland, E., Castegnaro, M.
A propos de l'action cancérigène de la gamma-butyrolactone chez les souris
C.R. Acad. Sc. Paris, Série D, 282, 799–802 (1976)

Salamone, M.F., Heddle, J.A., Katz, M.
Mutagenic activity of 41 compounds in the in vivo micronucleus test
Prog. Mutat. Res., 1, 686–697 (1981)

Schoental, R.
Pathological lesions, including tumors, in rats after 4,4-diaminodiphenylmethane and gamma-butyrolactone
Israel J. Medic. Sci., 4, 1146–1158 (1968)

Searle, C.E. (ed.)
Chemical Carcinogens, p. 190 (1976)
In: National Library of Medicine's Hazardous Substances Data Bank, on-line search 5/28/87

Sharp, D.C., Parry, J.M.
Induction of mitotic gene conversion by 41 coded compounds using the yeast culture JD1
Prog. Mutat. Res., 1, 491–501 (1981a)

Sharp, D.C, Parry, J.M.
Use of repair-deficient strains of yeast to assay the activity of 40 coded compounds
Prog. Mutat. Res., 1, 502–516 (1981b)

Simmon, V.F., Shepherd, G.F.
Mutagenic activity of 42 coded compounds in the Salmonella/microsome assay
Prog. Mutat. Res., 1, 333–342 (1981)

Skopek, T.R., Andon, B.M., Kaden, D.A., Thilly, W.G.
Mutagenic activity of 42 coded compounds using 8-azaguanine resistance as a genetic marker in Salmonella typhimurium
Prog. Mutat. Res., 1, 371–375 (1981)

Styles, J.A.
Activity of 42 coded compounds in the BHK-21 cell transformation test
Prog. Mutat. Res., 1, 638–646 (1981)

Swern, D., Wieder, R., McDonough, M., Meranze, D.R., Shimikin, M.B.
Investigation of fatty acids and derivatives for carcinogenic activity (1970)
In: IARC Monographs on the Evaluation of Carcinogenic Risk of Chemicals to Man, 11, 231–239 (1976)

Thomson, J.A.
Mutagenic activity of 42 coded compounds in the lambda induction assay
Prog. Mutat. Res., 1, 224–235 (1981)

Topham, J.C.
Evaluation of some chemicals by the sperm morphology assay
Prog. Mutat. Res., 1, 718–720 (1981)

Trueman, R.W.
Activity of 42 coded compounds in the Salmonella reverse mutation test
Prog. Mutat. Res., 1, 343–350 (1981)

Tsuchimoto, T., Matter, B.E.
Activity of coded compounds in the micronucleus test
Prog. Mutat. Res., 1, 705–711 (1981)

Tweats, D.J.
Activity of 42 coded compounds in a different killing test using E. coli strains WP2, WP67, (uvrA/polA) and CM871 (uvrA lex A recA)
Prog. Mutat. Res., 1, 199–209 (1981)

Ullmanns Enzyklopädie der technischen Chemie
4th Edition, Vol. 9, p. 48–50
Verlag Chemie, Weinheim (1975)

Van Duuren, B.L., Nelson, N., Orris, L., Palmes, E.D., Schmitt, F.L.
Carcinogenicity of epoxides, lactones and peroxy compounds
J. Natl. Cancer Inst., 31, 41–55 (1963)

Van Duuren, B.L., Orris, L., Nelson N.
Carcinogenicity of epoxides, lactones and peroxy compounds, II
J. Natl. Cancer Inst., 35, 707–717 (1965)

Venitt, S., Crofton-Sleigh, C.
Mutagenicity of 42 coded compounds in a bacterial assay using E. coli and S. typhimurium
Prog. Mutat. Res., 1, 351–360 (1981)

Vogel, E., Blijleven, W.G.H., Kortselius, M.J.H., Zijlstra, J.A.
Mutagenic activity of 17 coded compounds in the sex-linked recessive lethal test in Drosophila melanogaster
Prog. Mutat. Res., 1, 660–665 (1981)

Walkenstein, S.S., Wiser, R., Gudmunsden, C., Kimmel, H.
Metabolism of gamma-hydroxybutyric acid
Biochim. Biophys. Acta, 86, 640–642 (1964)

Zimmermann, F.K., Scheel, J.
Induction of mitotic gene conversion in strain D7 of Saccharomyces cerevisiae by 42 coded chemicals
Prog. Mutat. Res., 1, 481–490 (1981)

5-Nitroanisidine

1. Summary and assessment

An assessment of the acute toxic properties of 5-nitro-o-anisidine is scarcely possible on the basis of the available reports. In a sub-chronic experiment (7 weeks or more), administration of this substance in relatively high dosage in the feed of rats and mice results in loss of weight, liver damage, changes in the thyroid gland and spleen, and testicular damage in males.

In the Ames test using strain TA98, 5-nitro-o-anisidine produces a dose-related mutagenic effect, in the absence of a metabolic activating system. In vitro application to V79 cells of the Chinese hamster does not produce chromosome aberrations. No influence on DNA repair (tested on rat liver cells in vitro) can be demonstrated.

Chronic administration of 5-nitro-o-anisidine in the feed results in a dose-related increase in tumours of the breast, skin, cymbal's gland and clitoris. On the basis of the available findings, a significant carcinogenic potential must be assumed for this substance.

2. Name of substance

2.1	Common name	5-Nitroanisidine
2.2	IUPAC-name	2-Amino-1-methoxy-4-nitro-benzene
2.3	CAS-No.	99-59-2

3. Synonyms

Common and trade names

2-Methoxy-5-nitroaniline
5-Nitro-ortho-anisidine
Azoic Diazo Component 13, Base
Azoamine Scarlet K
2-Methoxy-5-nitrobenzen-amine

4. Structural and molecular formulae

4.1 Structural formula

4.2 Molecular formula $C_7H_8N_2O_3$

5. Physical and chemical properties

5.1	Molecular mass	168.16 g/mol
5.2	Melting point	118° C
5.3	Boiling point	no information available
5.4	Density	1.2068 g/cm^3 (20° C)
5.5	Vapour pressure	no information available
5.6	Solubility in water	soluble in hot water
5.7	Solubility in organic solvents	alcohol, ether, acetone, benzene
5.8	Solubility in fat	no information available
5.9	pH-value	no information available
5.10	Conversion factors	1 ppm $\hat{=}$ 6.87 mg/m^3 1 mg/m^3 $\hat{=}$ 0.146 ppm (Weast, 1981/82; IARC, 1982)

6. Uses

Intermediates for dyestuffs (IARC, 1982).

7. Results of experiments

7.1 Toxicokinetics and metabolism
No information available.

7.2 Acute and subacute toxicity
According to Lewis and Tatken (cited in IARC, 1978), the oral LD$_{50}$ of 5-nitroanisidine for rats is given as 704 mg/kg (no other information). Vasilenko and Swesdai(1981) determined an oral LD$_{50}$ of 2225 mg/kg for rats. For mice the oral LD$_{50}$ is given as 1060 mg/kg (cited in IARC, 1978, without further details).

The National Cancer Institute (1978) conducted a study of the subacute toxicity of 5-nitroanisidine for male and female B6C3F1 mice and male and female Fischer-344 rats. The dose groups and the control each comprised 5 males and 5 females of each species. The test substance was administered in the feed at concentrations of 0.05, 0.1, 0.2 and 0.4% for 7 weeks. The animals then received normal feed for one week. After 8 weeks all the surviving animals were killed and examined macroscopically. None of the rats died before the end of the test. Dark colouration of the spleen was observed in 2 female rats in the high dose group. According to the authors, the rats in the high dose group exhibited a reduction in the mean body weight of 0.7% in the males and of 10% in the females. None of the mice died during the test. A concentration of 0.4% 5-nitroanisidine in the feed produced only a decrease in the mean body weight of male mice of 7.9% compared with the control.

Administration of 0.8% 5-nitroanisidine in the feed over a period of 7 weeks caused, in male rats, degeneration of the liver and testes and sclerosis of the gastric mucosa, and in both male and female rats, pigmentation of the thyroid follicles and haematopoiesis in the spleen (National Cancer Institute, 1978).

7.3 Skin and mucous membrane effects
No information available.

7.4 Sensitization
No information available.

7.5 Subchronic/chronic toxicity
When 5-nitroanisidine was administered in the feed for 15 weeks at a concentration of 0.8%, there was weight reduction in mice; 1.6% in the feed caused degenerative hyperplastic liver lesions and inflammatory lymphocytic infiltration of the thyroid in male mice and myelofibrosis of the bone in female mice (National Cancer Institute, 1978).

7.6 Genotoxicity
Chiu et al. (1978) investigated 5-nitroanisidine in the Ames test for possible mutagenic action. The strains used were TA 98 and TA 100. No metabolizing system was added. Doses of 0.1, 1.0 and 10 μmol/plate were used. The test substance caused dose-related reversions in Salmonella typhimurium strain TA 98.

5-Nitroanisidine was tested in vitro on V 79 cells of the Chinese hamster for possible chromosome aberrations. The investigation was conducted with and without a metabolizing system (S9-mix, rat liver homogenate, Aroclor-induced). Doses of 250, 500 and 750 µg/ml were used. No chromosome aberrations were induced in either experiment (Müller, 1988).

5-Nitroanisidine was tested in primary rat hepatocyte cultures for possible induction of DNA repair synthesis. Doses from 0.03 to 3mM of the substance were used in two independent tests. In neither case were there indications of induction of DNA repair synthesis (Andrae, 1988).

7.7 Carcinogenicity

The National Cancer Institute (1978) tested the carcinogenic action of 5-nitroanisidine in a chronic feeding test in Fischer-344 rats and B6C3F1 mice, which were about 6 weeks old at the beginning of the test. The test arrangement is shown in Tables 1 and 2.

Table 1. Test arrangement for verifying the carcinogenic effect of 5-nitroanisidine after chronic administration in the feed of rats

	Number of animals per group	Concentration (%) of 5-nitro-anisidine	Observation time (weeks) treated	untreated
a) Male test animals				
Control A[a]	50	–	–	108
Control B[b]	49	–	–	109
Dose A	50	0.4	78	28
Dose B	50	0.8	78	24
b) Female test animals				
Control A[a]	50	–	–	108
Control B[b]	50	–	–	109
Dose A	50	0.4	78	28
Dose B	50	0.8	78	24

[a] Control used initially
[b] Control used subsequently

Table 2. Test arrangement for verifying the carcinogenic effect of 5-nitroanisidine after chronic administration in the feed of mice

	Number of animals per group	Concentration (%) of 5-nitroanisidine	Observation time (weeks) treated	Observation time (weeks) untreated
a) Male test animals				
Control A[a]	50	–	–	95
Control B[b]	50	–	–	96
Dose A	50	0.4	78	18
Dose B	50	1.6[c]	15	
		0.4	63	18
b) Female test animals				
Control A[a]	50	–	–	95
Control B[b]	50	–	–	96
Dose A	50	0.8	78	19
Dose B	50	1.6[c]	15	
		0.4	63	18

[a] Control used initially
[b] Control used subsequently
[c] Dose was reduced after 15 weeks because of toxicity of the substance

Initially the concentrations of 5-nitroanisidine administered in the feed were 0.2 and 0.4% for rats and 0.4 and 0.8% for mice. The rats and mice in the low dose groups (0.2 and 0.4% respectively) were killed after 16 weeks, because according to the authors these doses were regarded as too low. 7 months after the actual start of the study, a new group of rats (50 male, 50 female) received 0.8% 5-nitroanisidine in the feed, and a new group of mice (50 male, 50 female) received 1.6% of the test substance in the feed, and 2 new control groups were set up. In the mice, however, the concentration of 1.6% was lowered to 0.4% after 15 weeks. 49 and 78 weeks after the start of the test, 5 male und 5 female mice of each treated and control group were killed, and in the case of the rats, 5 male and 5 female animals of control group A were killed after 78 weeks and of control group B after 80 weeks, and examined macroscopically.

Table 3. Carcinogenic effect of 5-nitroanisidine in rats after chronic administration in the feed

a) Male test animals

Dose group	Cutaneous and associated glands[a]			Cymbal gland or skin of ear			Preputial gland		
	N	%	p-value[b]	N	%	p-value[b]	N	%	p-value[b]
Control A (0%)	1	2		1	2		2	4	
Control B (0%)	0	0		0	0		0	0	
Dose A (0.4%)	30	60	<0.001	2	4		2	4	
Dose B (0.8%)	40	83	<0.001	10	21	0.001	5	10	0.028

b) Female test animals

Dose group	Mammary glands[c]			Cymbal gland or skin of ear			Clitoris adenomas or carcinomas		
	N	%	p-value[b]	N	%	p-value[b]	N	%	p-value[b]
Control A (0%)	0	0		0	0		1	2	
Control B (0%)	2	4		0	0		2	4	
Dose A (0.4%)	11	22	0.007	3	6		12	25	0.001
Dose B (0.8%)	4	9	0.049	7	15	0.004	14	30	0.001

[a] Epithelial-cell and basal-cell carcinomas, trichoepitheliomas, adenocarcinomas of the sebaceous glands, carcinomas of the sweat glands
[b] Only stated if there was a higher incidence than in the corresponding control (method of Cox and Tarone)
[c] Adenocarcinomas, adenomas

Table 4. Carcinogenic effect of 5-nitroanisidine in mice after chronic administration in the feed

Dose group	Hepatocellular carcinomas		
	N	%	p-value[a]
a) Male test animals			
Control A (0%)	12	24	
Control B (0%)	6	13	
Dose A (0.8%)	25	52	0.004
Dose B (1.6–0.4%)	3	6	
b) Female test animals			
Control A (0%)	2	4	
Control B (0%)	1	2	
Dose A (0.8%)	0	0	
Dose B (1.6–0.4%)	8	19	0.008

[a] Only stated if the incidence was significantly higher than in the corresponding control (method of Cox and Tarone)

The female rats treated with 5-nitroanisidine exhibited significantly higher mortality than the control animals. In the case of the males, in all test groups more than 78% survived to the 70th test week. For the females, on the other hand, 88% of control group A, 86% of control B, 74% of dose A (0.4%) and 58% of dose B (0.8%) reached the 85th test week. At the end of the test, in the case of male mice, 86% of control group A, 80% of control group B, 84% of dose group A (0.8%) and 88% of dose group B (1.6–0.4%) were alive; in the case of the female mice, 72% of control group A, 76% of control group B, 32% of dose group A (0.8%) and 66% of dose group B (1.6–0.4%) were alive. The results of the carcinogenicity tests with 5-nitroanisidine are summarized in Tables 3 and 4. It should be noted at this point that in the case of the rats, the control and treated animals were from different shipments, and the rats in dose group A were not from the same supplier as the relevant control. Therefore it is not certain that all differences in the incidence of tumours between the rats of dose group A and their control can be attributed to administration of the test substance.

Similarly for the mice, the control and treated mice were from different shipments and the mice in dose group A and the associated control group were from different suppliers.

In addition, it is important to note that the incidences of liver tumours in the treated mice were comparable in order of magnitude to those that were observed in historical controls. For these reasons, in the opinion of the IARC Working Group, the experiment on mice cannot be evaluated (IARC, 1982).

Tumours of organs other than those listed in the tables were classified as unrelated to treatment, both in the rat and in the mouse.

7.8 Reproductive toxicity
No information available.

7.9 Effects on the immune system
No information available.

7.10 Other effects
No information available.

8. Experience in humans

No information available.

References

Andrae, U.
5-Nitro-o-anisidine
Test for the induction of DNA-repair in rat hepatocyte primary cultures
Gesellschaft für Strahlen- und Umweltforschung, study no. HeRe1–87/BGCH (1988)
Commissioned by the Employment Accident Insurance Fund of the Chemical Industry

Chiu, C.W., Lee, L.H., Wang, C.Y., Bryan, G.T.
Mutagenicity of some commercially available nitro compounds for Salmonella typhimurium
Mutat. Res., 58, 11 (1978)

Hoechst AG
5-Nitroanisidin
Chromosome aberrations in vitro in V79 chinese hamster cells

Study no. 87.0575 (1988) commissioned by the
Employment Accident Insurance Fund of the Chemical Industry

IARC (International Agency For Research On Cancer
Monographs on the evaluation of the carcinogenic risk of chemicals to humans
Vol. 27, 133–139 (1982)

National Cancer Institute
Bioassay of 5-Nitro-o-Anisidine for possible carcinogenicity
Bethesda, MD, US (1978)

Vasilenko, N.M., Swesdai, V.J.
Possibility of mathematical prediction of some criteria of toxicity of nitro and amino benzenoid compounds
Gig. Tr. Prof. Zabol., 8, 50–52 (1981)

Weast, R.C. (ed.)
Handbook of Chemistry and Physics,
62nd Edition, C-98
CRC-Press, Boca Raton, Florida, 1981/1982

N,N'-Di-sec.-butyl-p-phenylenediamine

1. Summary and assessment _____

In the following investigations, N,N'-di-sec.-butyl-p-phenylenediamine was used in the form of the technical products Kerobit BPD and Gasoline Antioxidant No. 22.

The oral LD_{50} in the rat has been reported as 121–178 mg/kg or 450 mg/kg.

Loss of weight, dyspnoea and slight liver damage are observed on repeated oral administration (14 days, 90 mg/kg ($^1/_5$ of the LD_{50})); however, these effects are reversible. A concentration of 0.01% in the feed (approx. 6 mg/kg/day) over three months is tolerated without toxic symptoms, histopathological changes or pathological alterations in the haematological or clinical chemistry parameters. At higher concentrations (0.05%, approx. 30 mg/kg/day) only loss of weight is observed, with no other changes in the parameters specified above.

N,N'-di-sec.-butyl-p-phenylenediamine (Kerobit BPD/Antioxidant No. 22) has an acute irritative or caustic effect on the skin and mucosa; it is absorbed through the skin. Following absorption (orally or dermally), methaemoglobin is formed, giving typical symptoms of poisoning. The technical product has a sensitizing effect on the skin.

In the Salmonella/microsome test on strains TA 1535, TA 1537, TA 1538, TA 100 and TA 98, Kerobit BPD, shows a weak mutagenic effect only on strain TA 100 with S9-mix at the highest tolerable concentration (the mutagenic rate being only 1.4–1.9 times the control value). In the HGPRT test on V79 cells and in the chromosome aberration test on V79 cells there is no mutagenic effect.

The BG Chemie has commissioned further experiments on repeated application of Kerobit BPD. This proved necessary because of the difficulty in interpreting the results of early subchronic feeding experiments, in which there was a possibility of low levels of antioxidants in the feed. The experimental standards at that time (1966) do not meet present-day requirements.

No methaemoglobin formation has been observed in workers involved in the production and handling of Kerobit BPD.

2. Name of substance

2.1 Usual name — N,N′-Di-sec.-butyl-p-phenylenediamine

2.2 IUPAC-name — N,N′-Bis(1-methylpropyl)-1,4-diaminobenzene

2.3 CAS-No. — 101-96-2

3. Synonyms

Common and trade names
N,N′-Bis(1-methylpropyl)-1,4-benzenediamine
Kerobit BPD
Tenamene 2
DuPont Gasoline Antioxidant No. 22

4. Structural and molecular formulae

4.1 Structural formula

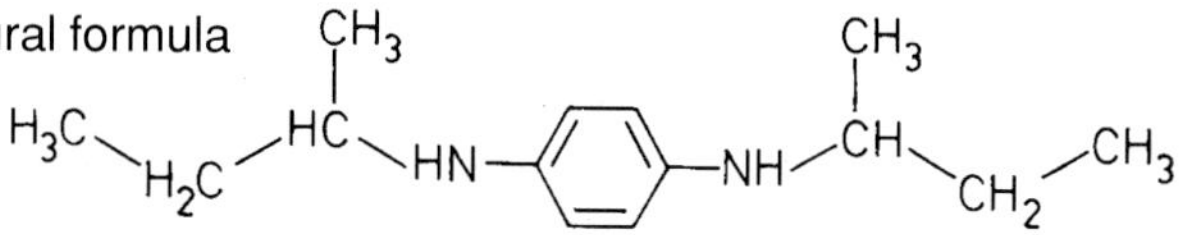

$H_3C-CH_2-CH(CH_3)-HN-C_6H_4-NH-CH(CH_3)-CH_2-CH_3$

4.2 Molecular formula — $C_{14}H_{24}N_2$

5. Physical and chemical properties

5.1 Molecular mass — 220.4 g/mol

5.2 Melting point — 14–18° C / 20° C

5.3 Boiling point — ca. 310° C (industrial product) / 296° C

5.4 Density — 0.942 g/cm^3 (at 20° C)

5.5 Vapour pressure — 1.6 hPa (at 20° C) / 44 hPa (at 196° C)

5.6 Solubility in water — very slight
Solubility of the hydrochloride — no information available

5.7 Solubility in organic solvents — soluble in alcohol, DMSO, petrol

5.8	Solubility in fat	no information available
5.9	pH-value	9 (at 1 g/l water)
5.10	Conversion factors	1 ppm $\hat{=}$ 9.145 mg/m^3 1 mg/m^3 $\hat{=}$ 0.109 ppm (BASF, 1980; Du Pont de Nemours, 1966)

6. Uses

Antioxidant for prolonging the storage life of petrol (Römpps, 1983). Antioxidant in carrier oils of plant protection agents (Downer and Phillips, 1978).

7. Results of experiments

7.1 Toxicokinetics and metabolism
No information available.

7.2 Acute and subacute toxicity
The following results have been obtained in acute toxicity tests:

Table 1. Acute toxicity of Kerobit BPD

Species	Route of adminis- tration	LD$_{50}$ mg/kg	Symptoms	Dissection
Rat (Wistar)	Oral	>121 <178	Staggering Dyspnoea Poor GC[a]	Isolated fatty change in liver cells (BASF, 1982/ 1984)
Rat	Oral	450	Staggering Pallor Weight loss	Liver and kid- ney damage (Du Pont, 1966)
Rat (Wistar)	i.p.	>10 <50	Staggering Dyspnoea Poor GC[a]	Intraabdominal adhesions (BASF, 1982/ 1984)

[a] General condition

Table 1 (continued)

Species	Route of administration	LD$_{50}$ mg/kg	Symptoms	Dissection
Rat (Wistar)	Dermal (24 hours occlusive)	756 (623–903)	Local oedema Skin necroses otherwise normal	Deep-seated skin necroses isolated fatty changes in liver cells (BASF, 1982/1984)
Rat (Charles River)	Dermal (rubbed-in, uncovered)	ca. 450	Cyanosis Weakness Weight loss Met-Hb after 5 hours: Control: 0.18 g/100 ml 450 mg/kg: 1.91 g/100 ml 1500 mg/kg: 2.75 g/100 ml	Skin necroses (Du Pont, 1966)
Cat	Oral	50 mg/kg 1/4 died Met-Hb Up to 23% (2 h) Up to 17% (4 h)	Salivation Cyanosis Lateral position	(BASF, 1982/1984)

Six male white rats received 90 mg/kg of industrial N,N'-di-sec.-butyl-p-phenylenediamine (Gasoline Antioxidant No. 22) orally 5 times weekly for 2 weeks. During this time there was weight loss, dyspnoea and excretion of grey coloured urine. Dissection and histological examination showed slight liver changes (no precise details). Both clinical signs and pathological changes were reversible (observations only on a total of 6 animals; Du Pont de Nemours, 1966).

90 mg/kg of Gasoline Antioxidant No. 22 was rubbed into the shaved skin of the back of 3 male white rats daily 5 times weekly for 2 weeks. Locally there were signs of irritation and ulceration. Symptoms of intoxication were not observed. In comparison with 3 control animals, the blood parameters were unchanged apart from granulocytosis in the treated animals; the Met-Hb level was not raised (Du Pont de Nemours, 1966).

7.3 Skin and mucous membrane irritation

Dermal application of undiluted Kerobit BPD to the back of rabbits and guinea pigs caused, after just 1 minute of action, formation of oedema at the application site, and black-brown necroses, which after incrustation healed with scarring in 3–4 weeks. Undiluted Kerobit BPD and a 50% oil solution applied to the skin of the back of white rabbits for 20 hours caused severe inflammation followed by severe black-brown necroses, which healed in 4–6 weeks with scarring. A 10% solution of Kerobit BPD caused slight inflammatory irritation with grey-red coloration, which subsided after about 7 days, with scale formation (BASF, 1982a).

In the rabbit eye, instillation of one drop of undiluted Kerobit BPD caused severe oedema, purulent secretion from the conjunctivae, necroses and incrustation at the margins of the eyelids. Scarring remained after healing. There was temporary corneal clouding, but permanent corneal lesions were not observed (BASF, 1982a).

7.4 Sensitization

The sensitizing effect of Kerobit BPD (purity 96.9%) was tested in the open epicutaneous test (Klecak, 1977) on female guinea pigs (Pirbright White breed). In a preliminary test, 10%, 3%, 1% and 0.3% Kerobit solutions in ethanol were applied (uncovered) on the shaved skin of the flank. The minimum irritant concentration was found to be a 1% preparation, and the maximum non-irritant concentration was found to be a 0.3% preparation of the substance. The following concentrations were chosen for induction in the main test: 3%, 1%, 0.3%, 0.1%, 0.01% and 0.005% in ethanol, with enthanol alone as control. Groups of 8 animals received applications of 0.1 ml/8 cm^2, 5 times weekly (total of 20 times) to the shaved right flank. The effects on the skin were recorded 24 hours after each application. 28 days and 42 days after the first application, 0.025 ml/2 cm^2 of the various Kerobit solutions was applied to the left flank, and the findings were recorded after 24, 48 and 72 hours. The 3%, 1%, 0.3% and 0.1% preparations of the substance produced definite inflammatory reactions on the left flank, whereas the 0.01 and 0.005% concentrations did not cause any skin changes. Thus, Kerobit BPD has a sensitizing effect on the skin of the guinea pig even at concentrations of 0.1% (BASF, 1982/1984).

7.5 Subchronic/chronic toxicity

Groups of six male and 6 female newly-weaned albino rats received 0, 0.002, 0.01 and 0.05% of Gasoline Antioxidant No. 22 in the feed (20, 100, 500 ppm) for 90 days. In the case of the 0.002% dose group, the concentration in the feed was increased after 65 days to 0.1% for 25 days and then to 0.2% for 4 days. In the 0.05% group, the males and, to a lesser extent, the females exhibited an increasing delay in growth. In the 0.002% group when the concentration of gasoline antioxidant was increased, there was a definite weight loss. 0.01% of gasoline antioxidant in the feed did not cause any weight changes in comparison with the controls. None of the test groups showed symptoms of intoxication, pathological-histological organ changes (organs not specified), impairment of the blood picture or pathological changes of bilirubin and alkaline phosphatase in the plasma. The Met-Hb was not raised (Du Pont de Nemours, 1966).

7.6 Genotoxicity

In vitro. Kerobit BPD (purity 98.5%) was tested in the Salmonella/microsome test on strains TA 1535, TA 1537, TA 1538, TA 100 and TA 98 with and without metabolic activation (S9-mix from Aroclor-induced rat liver). Concentrations of 20 to 5000 µg/plate were used, and a toxic efffect on the bacteria was observed starting from 500 µg/plate. Without metabolic activation,Kerobit BPD was not mutagenic. With metabolic activation, a slight, dose-dependent mutagenic effect was observed for strain TA 100 at concentrations of 250 and 500 µg/plate. However, the maximum increase of the mutation rate for TA 100 was only 1.4 to 1.9 times the control value. It cannot be excluded that this weak mutagenic effect was caused by contaminants in the industrial product (BASF, 1982b).

In the HGPRT test with V79 cells, at concentrations of 7.5 to 75 ng/ml without S9-mix and 1.65 to 8.5 µg/ml with S9-mix, in each case incubated for 4 hours, there were no indications of mutagenic activity. At the highest concentrations the cell survival rates were 35.1 and

27.7% without S9-mix and 45.3 and 29.7% with S9-mix (two experiments in each case; Miltenburger, 1987a).

In an in vitro chromosome aberration test on V79 cells of the Chinese hamster, no increased rate of chromosome aberrations was found, with or without S9-mix; the concentrations tested were 3 to 75 ng/ml in the tests without S9-mix, and 1.3 to 8.5 µg/ml with S9-mix (Miltenburger, 1987b).

In vivo. No information available.

7.7 Carcinogenicity
No information available.

7.8 Reproductive toxicity
No information available.

7.9 Effects on the immune system
No information available.

7.10 Other effects
No information available.

8. Experience in humans

No methaemoglobin formation was observed in workers involved in the manufacture and handling of Kerobit BPD (BASF, 1988).

References

BASF AG
Safety data sheet "Kerobit BPD", 1980

BASF AG
Communiaction from BASF AG dated 10.5.1982 to the Employment Accident Insurance Fund of the Chemical Industry (1982a)

BASF AG, Abteilung Toxikologie
Report on the testing of Kerobit BPD in the Ames test
Unpublished Report (1982b)

BASF AG, Abteilung Toxikologie
Toxicological investigation of Kerobit BPD
Unpublished Report 1982/1984

BASF AG
Written communication from the department of industrial medicine and protection of health, dated 1.2.1988

Downer, J.D., Phillips, C.A.
Agricultural spray oil containing oxidation inhibitors
U.S. Patent No. 4,125,400 dated 14.11.1978

E.I. Du Pont de Nemours
Data sheet DS-66-001, 4-28-66
Gasoline Antioxidant No. 22 (N,N'-Di-(sec.-butyl)-p-phenylenediamine)

Klecak, G.
Identification of contact allergens
Advances in Modern Toxicology, Vol. 4, 305–339 (1977)

Miltenburger, H.G.
Detection of gene mutations in somatic mammalian cells in culture: HGPRT-test with V79 cells
Laboratory for Mutagenicity Testing, Darmstadt, 6.7.1987
Commissioned by the Employment Accident Insurance Fund of the Chemical Industry (1987a)

Miltenburger, H.G.
Chromosome aberrations in cells of Chinese hamster cell line V79
Laboratory for Mutagenicity Testing, Darmstadt, 3.6.1987
Commissioned by the Employment Accident Insurance Fund of the Chemical Industry (1987b)

Römpps Chemie Lexikon
Vol. 3, p. 2094, 8th Edition
Franckh'sche Verlagshandlung, Stuttgart (1983)

m-Cresidine

1. Summary and assessment

No studies on the toxicokinetics or acute toxicity of m-cresidine have so far been reported.

In a study of the subacute effects over a seven-week period, doses of between 0.15 and 0.3 g m-cresidine/kg/day caused a reduction in body weight in rats and mice; no further toxic damage is reported.

Long-term administration (2 years or 18 months) leads to chronic toxic damage to the kidneys and renal tracts.

In the Salmonella/microsome test and in in-vitro testing with Escherichia coli, m-cresidine gives negative results with and without metabolic activation.

In a study of carcinogenicity using technical grade m-cresidine, a dose-related increase in bladder tumours is found in exposed male and female rats but not in the control animals. The increase is not, however, statistically significant. In female mice, a dose-related increase in liver tumours is observed; the survival time in male animals is too short for evaluation. m-Cresidine is positive in a cell transformation test on embryonic rat cells infected with Rauscher leukaemia virus.

The available experimental findings indicate a possible carcinogenic effect of m-cresidine, but they are not statistically significant.

2. Name of substance

2.1 Common name   m-Cresidine
2.2 IUPAC-name   1-Amino-4-methoxy-2-methylbenzene
2.3 CAS-No.   102-50-1

3. Synonyms

Common and trade names   2-Methyl-p-anisidine
                                4-Methoxy-2-methylaniline

m-Cresidine

2-Methyl-4-methoxyaniline
4-Methoxy-2-methyl-
   benzenamine

4. Structural and molecular formulae

4.1 Structural formula

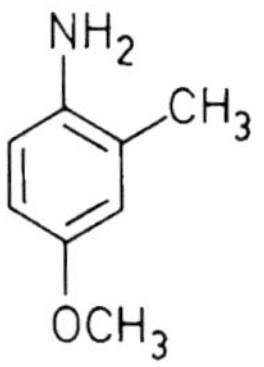

4.2 Molecular formula $C_8H_{11}NO$

5. Physical and chemical properties

5.1	Molecular mass	137.2 g/mol
5.2	Melting point	29–30° C
5.3	Boiling point	248–249° C
5.4	Density in g/cm^3	no information available
5.5	Vapour pressure, hPa	no information available
5.6	Solubility in water	no information available
5.7	Solubility in organic solvents	soluble in ethanol, petroleum ether (hot)
5.8	Solubility in fat	no information available
5.9	pH-value	no information available
5.10	Conversion factors	1 ppm $\hat{=}$ 5.6 mg/m^3 1 mg/m^3 $\hat{=}$ 0.179 ppm (IARC, 1982)

6. Uses

Starting product for azo dyes (IARC, 1982).

7. Results of experiments

7.1 Toxicokinetics and metabolism
No information available.

7.2 Acute and subacute toxicity

No studies are available relating to the acute toxicity of m-cresidine.

A subacute study was conducted in 1978 by the US National Cancer Institute on Fischer-344 rats and B6C3F1 mice, which were administered m-cresidine in corn oil, by stomach tube. Six groups each of 5 male and 5 female rats and mice received doses of 0.02, 0.04, 0.08, 0.15 and 0.3 g/kg/day (rats) and 0.04, 0.08, 0.15, 0.3 and 0.6 g/kg/day (mice). There was one control group for each species. Treatment was given on 5 days per week for 7 weeks with a follow-up period of 1 week. At the dose of 0.15 g/kg/day, one female rat died. The average weight reduction was 3.6% (0.08 g/kg/day), 4.5% (0.15 g/kg/day) and 26.5% (0.3 g/kg/day) for the male rats, and 6.5% (0.3 g/kg/day) for the females. In mice, deaths occurred amongst the males given 0.3 and 0.6 g/kg/day and the females given 0.08 g/kg/day or more. The average weight reduction was given as 13.4% (0.15 g/kg/day) and 8.4% (0.3 g/kg/day) for the male mice and 11.1% (0.3 g/kg/day) and 18.4% (0.6 g/kg/day) for the females.

7.3 Skin and mucous membrane effects

No information available.

7.4 Sensitization

No information available.

7.5 Subchronic/chronic toxicity

Long-term administration (2 years or 18 months) led to chronic toxic damage to the kidneys and renal tracts, thymus, spleen and testes (for further details see 7.7)

7.6 Genotoxicity

Rosenkranz and McCoy (cited in IARC, 1982) found no evidence of mutagenic activity of m-cresidine (the sample corresponded to that used in the carcinogenicity test) with Escherichia coli WP2uvrA and Salmonella typhimurium TA 1537, TA 1538, TA 98 and TA 100. Tests were conducted with and without a metabolizing system (Aroclor-induced and uninduced liver microsomes of mouse, rat and hamster).

7.7 Carcinogenicity

In 1978 the US National Cancer Institute conducted an investigation of carcinogenesis with m-cresidine. The substance was administered in corn oil to Fischer-344 rats and B6C3F1 mice, by stomach tube. The test arrangement is shown in the following tables.

Table 1. Test arrangement of the study for verifying the carcinogenic effect of m-cresidine after chronic oral administration to rats[a]

	Number of animals per group	Dose[b] (g/kg/day)	Test duration (weeks) Treated	Test duration (weeks) Untreated	Tumours[c] (%)
a) Male test animals					
Untreated	50	–	–	110	0
Solvent-Control	25	–	–	110	0
Dose A	50	0.08	77	32	0
Dose B	50	0.16	77	33	11
b) Female test animals					
Untreated	50	–	–	110	0
Solvent-Control	25	–	–	110	0
Dose A	49	0.08	77	32	2
Dose B	50	0.16	77	33	5

[a] Ages of animals at start of test: 6 weeks (solvent control, dose A and dose B), 10 weeks (untreated control)
[b] 5 days per week by stomach tube
[c] Papillary transitional-cell carcinomas

For the mice, the original doses were reduced after 32 weeks because of high mortality in the treated groups. All test animals were weighed twice weekly for the first 12 weeks, and then at monthly intervals. On dissection, all animals were assessed macroscopically, and the following organs were examined histologically: skin, subcutaneous tissue, lungs, bronchi, trachea, bone marrow, spleen, lymph nodes, thymus, heart, salivary glands, liver, gall bladder (mouse), pancreas, oesophagus, stomach, small intestine, large intestine, kidney, urinary bladder, hypophysis, thyroid, parathyroid, brain, testes, prostate, uterus, mammary glands and ovaries.

In the male rats of the high-dose group (0.16 g/kg) a reduced weight gain was observed during the first 48 weeks, but after this time weight gain was similar to that of the controls. Growth of the males in the low-dose groups and all treated females was comparable to the controls. Mortality among the treated male rats showed

Table 2. Test arrangement of the study for verifying the carcinogenic effect of m-cresidine after chronic oral administration to mice [a]

	Number of animals per group	Dose[b] (g/kg/day)	Test duration (weeks) Treated	Test duration (weeks) Untreated	Tumours[c] (%)
a) Male test animals					
Untreated	50	–	–	98	0
Solvent-Control	25	–	–	95	0
Dose A	50	0.08 0.02	32 21	40	0
Dose B	50	0.16 0.04	32 21	25	0
b) Female test animals					
Untreated	50	–	–	98	0
Solvent-Control	25	–	–	95	0
Dose A	50	0.08 0.02	32 21	41	2
Dose B	50	0.16 0.04	32 21	41	11.1

[a] Ages of animals at start of test: 6 weeks
[b] 5 days per week by stomach tube
[c] Hepatocellular adenomas and carcinomas

definite dependence on dose. 64% of the untreated control, 68% of the solvent control, 62% of the animals of dose A (0.08 g/kg) and 48% of the animals of dose B (0.16 g/kg) were still alive at the end of the test. The equivalent figures for the female rats were 72% of the untreated controls, 52% of the solvent control, 57% of the dose A animals (0.08 g/kg) and 44% of the dose B animals (0.16 g/kg). Carcinomas of the bladder (papillary transitional-cell carcinomas), considered to be treatment-related, were seen in 5 out of 44 (11%) of the male rats on the high dose (0.16 g/kg), in 1 out of 44 (2%) of the females on the low dose, and in 2 out of 44 (5%) of females on the high dose. The increased occurrence of tumours was not statistically significant, but was dose-dependent (for males p = 0.014). In historical controls at the same test laboratory, this type of bladder tumour had only occurred in this breed of rat in 1/249 females and 0/250 males, and the authors therefore considered the present result to be treatment-related. Pyelohyperplasia was found in 26 out of 45 male rats and 29 out of 45 female rats in the high-dose group, but not in the controls or the animals in the low-dose group (NCI, 1978). Other authors have supported or confirmed this opinion when classifying the substance (Griesemer and Cueto, 1980; Weisburger, 1983).

In the male mice there was a positive relation between dose and mortality (p <0.001). The median survival times were 52 weeks for males in the low-dose group and 26 weeks for those in the high-dose group (these survival times are inadequate for a carcinogenicity test). In the female mice, 80–84% of the controls, 86% of the low-dose group and 60% of the high-dose group (significantly increased mortality) were still alive at the end of the test. Treatment-related hepatocellular adenomas and carcinomas were observed in the females; 0 out of 23 in the solvent control, 1 out of 50 (1 carcinoma) in the low-dose group and 5 out of 45 (4 carcinomas) in the high-dose group. The incidence of hepatocellular neoplasms was increased in a dose-dependent manner (p = 0.027). Chronic-toxic kidney lesions (necroses of renal papillae and nephroses) and atrophy of the spleen and thymus (especially in male mice) were also found, and were dose-dependent. There was also a dose-dependent increase in the occurrence of multinuclear giant cells in the testes of the males (NCI, 1978).

An in-vitro test was conducted by Traul et al. (1981) with embryonic rat cells that had been infected with the Rauscher leukaemia virus (2 FR450). The indicator cells were treated with the test substance for 72 hours and then transferred to Petri dishes on agar. After 6 days the visible cell colonies of the treated cultures were compared with the control cultures. As a result of treatment with a carcinogen, there is loss of contact inhibition of the cells infected with Rauscher leukaemia virus, which is also manifested in an increased clonability in agar (increased cell survival rate in comparison with the controls). In this test m-cresidine was tested at doses of 51.5 and 83 µg per 5.2×10^4 cells, and gave a positive result.

7.8 Reproductive toxicity
No information available.

7.9 Effects on the immune system
No information available.

7.10 Other effects
No information available.

8. Experience in humans

No information available.

References

Griesemer, R.A., Cueto, C. Jr
Toward a classification scheme for degrees of experimental evidence for the carcinogenicity of chemicals for animals
IARC Sci. Publ., 27, 259–281 (1980)

IARC (International Agency for Research on Cancer)
Monographs on the evaluation of carcinogenic risk of chemicals to man 27, 91 (1982)

NCI (National Cancer Institut) Bethesda MD (U.S.)
Bioassay of m-cresidine for possible carcinogenicity
Technical Report Series No.105 (1978)

Traul, K.A., Takayama, K., Kachevsky, V., Hink, R.J., Wolff, J.S.
A rapid in vitro assay for carcinogenicity of chemical substances in mammalian cells utilizing an attachment-independence endpoint
J. Appl. Toxicol., 1, 190–195 (1981)

Weisburger, E.K.
Species differences in response to aromatic amines
Basic Life Sciences 27, 23–47 (1983)

2-Ethylhexanol

1. **Summary and assessment**

In experimental animals, 2-ethylhexanol is rapidly eliminated, mainly in the urine, partly unchanged and partly metabolized. The major metabolites in the urine have been identified as 2-ethylhexanoic acid, 2-ethylhexanoic acid glucuronide, 2-ethyl-1,6-hexanedioic acid and 2-ethyl-5-hydroxyhexanoic acid.

According to the available reports, 2-ethylhexanol is of relatively low acute toxicity in animals (oral LD_{50} in the rat is between 2049 and 5000 mg/kg, intraperitoneal LD_{50} is between 500 and 1000 mg/kg; dermal LD_{50} in the rabbit is between 1980 and >2600 mg/kg, in the rat >3000 mg/kg; LD_{50} in the rat on inhalation >230 ppm (approximates to the saturation level at room temperture) for 6 hours).

On repeated oral administration to mice and rats, 2-ethylhexanol results chiefly in liver damage (in particular peroxisome proliferation), hypolipidaemia and hepatomegaly. A "no observed effect level" has not yet been precisely determined, but for the rat it is probably above 350 mg/kg daily. It appears that similar pathological changes occur in rodents and dogs, but occur only to a limited extent, if at all, in humans and rhesus monkeys. No specific toxic effects are observed in rats following repeated dermal administration (1.67 g/kg daily for 12 days).

2-Ethylhexanol is moderately to strongly irritant when applied to the skin and eyes of experimental animals.

This substance shows no mutagenic activity in experiments on gene and chromosome mutation (in vitro or in vivo) or in the dominant lethal test. In only one experiment (Salmonella microsome test) there is a small increase in the number of revertants produced by 2-ethylhexanol. The cell transformation tests so far reported are negative.

A long-term study of oral administration to mice and rats, commissioned by the EPA, is now in preparation.

On oral application to rats, 2-ethylhexanol produces embryotoxic and teratogenic changes, but only at doses which show maternal toxicity (about half the LD_{50}). The effect of lower or borderline doses is potentiated by simultaneous administration of caffeine.

2-Ethylhexanol

No sensitizing properties of 2-ethylhexanol have been observed in humans, and there are no reports of such effects resulting in practice after contact with this substance.

2. Name of substance

2.1	Usual name	2-ethylhexanol
2.2	IUPAC-name	2-ethylhexan-1-ol
2.3	CAS-no.	104-76-7

3. Synonyms

Common and trade names

2-ethylhexanol-1
2-ethylhexyl alcohol
2-ethyl-n-hexyl-alcohol
octanol, technical
isoctanol
"AH"

4. Structural and molecular formulae

4.1 Structural formula

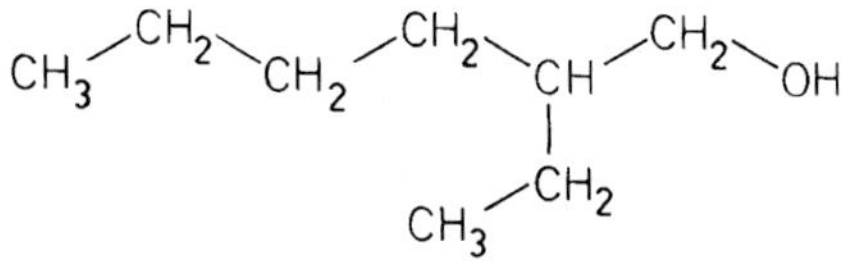

$CH_3-CH_2-CH_2-CH_2-CH(CH_2-CH_3)-CH_2-OH$

4.2 Molecular formula $\quad C_8H_{18}O$

5. Physical and chemical properties

5.1	Molecular mass	130.23 g/mol
5.2	Melting point	−76 to −70° C
5.3	Boiling point	183.5–185.0° C
5.4	Density	0.8329 g/cm^3 (at 20° C)
5.5	Vapour pressure	0.5 hPa (at 20° C) 13.3 hPa (at 78.7° C) 53.2 hPa (at 104.6° C)
5.6	Solubility in water	0.07 g/100 ml water (at 20° C)

5.7	Solubility in organic solvents	soluble in ether, ethanol and almost all organic solvents
5.8	Solubility in fat	no information available
5.9	pH-value	no information available
5.10	Conversion factors	1 ppm $\hat{=}$ 5.32 mg/m^3 1 mg/m^3 $\hat{=}$ 0.188 ppm (Ullmann, 1975, 1987; Clayton and Clayton, 1982; NAPIRI, 1974; Merck, 1983)
5.11	Odour	unpleasant
5.12	Odour threshold	
	(dissolved in water) at room temperature	1.28 ppm (range 0.58–2.08)
	at 60° C	0.78 ppm (range 0.58–124)
	no further details	0.27 ppm (Zoeteman et al., 1974; Hollenbach et al., 1972; Lillard and Powers, 1975).

6. Uses

Intermediate product in the manufacture of plasticisers, 2-ethylhexylacrylate, lubricant additives and surface treatment agents (Ullmann, 1987).

7. Results of experiments

7.1 Toxicokinetics and metabolism

2 male rats (Charles River, approx. 300 g) received 9 µg 2-ethyl [1-^{14}C] hexanol (about 30 µg/kg body weight), and 2 other males recieved 9 µg 2-ethyl [1-^{14}C] hexanol and 0.64 mM unlabelled 2-ethylhexanol (about 275 mg/kg body weight), administered by stomach tube. Elimination of radioactivity was almost complete after 28 hours. The recovered rate for the radioactivity was 96.1% (see the following overview). Distribution and elimination showed no differences at the high and low dosage levels.

Overview (% total radioactivity after 28 hours)

Faeces	8– 9
Urine	80–82
CO_2	6– 7
Carcass	1.4
Cage washings	2.7

The literature on the metabolism of 2-ethylhexanol yielded the data summarized in Table 1. The metabolites were determined exclusively in the urine. The chief metabolites in the urine were 2-ethylhexanoic acid, 2-ethylhexanoic acid glucuronide, 2-ethyl-1,6-hexanedioic acid and 2-ethyl-5-hydroxyhexanoic acid (Albro, 1975)

7.2 Acute and subacute toxicity

Following a single oral or intraperitoneal dose, the following LD_{50} values were established:

Mouse oral	between 2500 and 3830 mg/kg
Mouse i.p.	between 375 and 891 mg/kg
Rat oral	between 2049 and ca 5000 mg/kg
Rat i.p.	between 500 and 1000 mg/kg
Guinea pig oral	1860 mg/kg
Rabbit oral	1180 mg/kg

See also Table 2.

The following symptoms of poisoning in the mouse and the rat occurred: apathy, reeling, dyspnoea, cyanosis, incoordination, ataxia, reduced mobility, and prostration (Hodge, 1943; BASF, 1963; Scala and Burtis, 1973). The surviving animals recovered within 2 to 3 days of administration. Dissection revealed gastro-intestinal irritation following oral administration and also, following intraperitoneal injection, residues of the substance in the abdominal cavity and a "yellowish brown" liver (BASF, 1963; Scala and Burtis,1973).

Starting at a dose level of 0.20×10^{-5} mol/kg (about 0.26 mg/kg), intravenous administration of 2-ethylhexanol to rabbits led to a dose-dependent hypotonia. Following the injection of 3.2×10^{-5} mol/kg (about 4.16 mg/kg) there was a drop in blood pressure to 63%

of the original level. This reaction was accompanied by increased heart and respiratory rates. At the lower doses, there was rapid recovery from all these effects (Hollenbach et al., 1972).

In a dog, no clear hypotensive effects could be established at dose levels of 4.05 and 8.10×10^{-5} mol 2-ethylhexanol/kg (about 5.3 mg/kg and 10.5 mg/kg, respectively; Hollenbach et al., 1972).

Following a single dermal application to rats, rabbits and guinea pigs (duration of exposure 24 hours), the following LD_{50} values were obtained:

Rat	>3000 mg/kg
Rabbit (covered contact)	between 1980 and >2600 mg/kg
Guinea pig	>8300 mg/kg

See also Table 2.

In the case of the guinea pig, moderate skin irritation occurred during the experiment (no further details; Patty, 1982).

One dermal application of 2.6 g 2-ethylhexanol/kg for 24 hours to the shaved dorsal skin of rabbits (covered contact for 24 hours) did not produce symptoms of systemic poisoning. Moderate irritant effects developed at the application site (Scala and Burtis, 1973).

Rats similarly showed no substance-related symptoms of systemic poisoning following dermal application (Hüls, 1987).

In an inhalation-hazard test (20° C), an 8-hour exposure to 2-ethylhexanol produced no symptoms of poisoning in any of the rats used (6 or 12 animals) apart from escape movements at the beginning of the experiment (Smyth et al., 1969; BASF, 1963). No changes were shown up macroscopically at dissection (BASF, 1963).

Inhalation of 2-ethylhexanol at concentrations of 227 ppm or 235 ppm (atmosphere saturated at room temperature) for 6 hours did not lead to mortality in mice, rats (Wistar) or guinea pigs. The signs of poisoning were reduced motility, slight to moderate dyspnoea, and moderate irritation of the eyes, nose, pharynx and snout. The symptoms had subsided again in rats 1 hour after exposure. On dissection, macroscopically slight haemorrhages were found in the lungs (Patty, 1982; Scala and Burtis, 1973).

In the mouse, a concentration of 234.4 mg/m^3 air (about 44 ppm) reduced the respiratory rate by 50% ("FRD_{50}"; no further data; Muller and Greff, 1984).

There are a number of studies in the literature dealing with the effect of 2-ethylhexanol on the liver. Administration to male Swiss-Webster mice of 2-ethylhexanol for 10 days at a concentration of 2% in the diet (corresponding to ca. 2000 mg/kg body weight/day for an assumed food consumption level of 100 g/kg body weight/day) led to a significantly ($p \leq 0.05$) increased absolute liver weight, to significantly ($p \leq 0.05$) increased levels of cytoplasmic and microsomal epoxide hydrolase and gluthatione-S-transferase activity and also to a significantly ($p \leq 0.05$) increased cytoplasmic and microsomal protein content in the liver. The investigations were carried out on 6 control and 6 treated animals (Hammock and Ota, 1983).

Following daily administration by stomach tube of up to 13.5 mmol/kg (about 1758 mg/kg, dissolved in maize oil) to groups of 5 male and 5 female mice per dose for 14 days, toxic effects occurred at the highest dosage level (3 animals were killed in extremis). The relative liver weights were increased and there was peroxisome proliferation (increased palmitoyl-CoA activities). The effects were dose-dependent (in the brief communication, no information was given on dosage levels below the maximum dose) and were approximately half as severe on a molar basis as in the case of diethylhexyl adipate. Catalase activity remained unaffected. The same experiments were conducted on rats, with similar but more marked effects (Keith et al., 1985).

In related work, male rats (Alderly Park Wistar) received 1 mmol 2-ethylhexanol (about 130 mg/kg/day) for 14 days by stomach tube. In the 5 treated animals, no differences were noted in body weight, relative liver and testicular weights, number of peroxisomes, catalase activity in the liver, and the cholesterol and triglyceride levels, compared with 10 control animals. Histological examination of the liver failed to reveal any effects (Rhodes et al., 1984).

Sprague-Dawley rats (6 animals) received 2.7 mmol 2-ethylhexanol/kg/day for 5 days by stomach tube (about 351.6 mg/kg/day). In this study, an examination was made of the effect of 2-ethylhexanol on body weight and on testicular, prostate and liver weights. The seminiferous tubules were examined histologically. No indications were found of any changes (Sjöberg et al., 1986).

In a further study, groups of 10 rats received 330 or 660 mg 2-ethylhexanol/kg/day over a period of 17 days or 1320 mg/kg/day over a period of 22 days, in each case on 5 days a week (corresponding to 13 and 16 days of treatment). In both low-dose and high-dose groups there was significant ($p \leq 0.05$) reduction in body-weight gain.

Moreover, 2 animals died in the group receiving the high dose. No other parameters were examined in this study (Schmidt et al., 1983).

Oral administration of 1335 mg 2-ethylhexanol/kg/day for 7days to male Wistar rats (6 animals) led to a significantly ($p \leq 0.001$) increased relative liver weight and also to significantly ($p \leq 0.001$) raised cytochrome- P-450 and biphenyl-4-hydroxylase activity in the liver. The glucose-6- phosphatase activity was significantly lowered ($p \leq 0.01$). There were no changes in the activity of succinate dehydrogenase or aniline-4-hydroxylase, or in the microsomal protein content of the liver. Electron microscopy revealed a raised number of peroxisomes and an extended endoplasmic reticulum in the liver. Histochemical analysis showed an increase in the alcohol-dehydrogenase activity in the centrolobular zone of the liver (Lake et al., 1975).

In a further study on the rat (no details of the strain or the number of animals used), the administration of 2-ethylhexanol at a dietary level of 2% for 2 weeks (about 2000 mg/kg body weight/day for an assumed food consumption level of 100 g/kg body weight/day) had no subsequent effect on the activity of palmitoyl-CoA, catalase, uric acid oxidase, cytochrome-c oxidase, carnitine acetyl transferase, cytochrome P-450 or NADPH cytochrome c reductase, or on the protein content of the liver. Peroxisome proliferation was not observed in these studies, in contrast to the case of DEHP and MEHP (Ganning et al., 1982).

In further studies on the rat (F-344), 5 males received 2-ethyl-hexanol in the diet for 3 weeks. The concentration in the food was 2% (about 2000 mg/kg body weight/day for an assumed food consumption level of 100 g/kg body weight/day). A group of 13 animals served as the control. 2-Ethyl-hexanol led to a significantly ($p \leq 0.05$) reduced serum cholesterol level (40.0 ± 4.6 mg/100 ml; control 46.1 ± 4.8 mg/100 ml) and serum triglyceride level (59.2 ± 23.9 mg/100 ml; control 114.8 ± 17.8 mg/100 ml) and in the liver to significantly ($p \leq 0.01$) raised catalase activity (63 ± 5.2 units/mg protein; control 44 ± 2.7 units/mg protein) and carnitine acetyl transferase activity (40.1 ± 3.1 units/mg protein; control 2.7 ± 0.5 units/mg protein). At the end of the experiment there was a significantly ($p \leq 0.01$) incrased relative liver weight. There was also a change in the ratio of mitochondria to peroxisomes from 5:1 (control) to at least 1:1 or more (Moody and Reddy, 1976, 1982). These results conflict somewhat with Ganning et al. (1982), but this cannot be clarified at present.

10 rats (170–210 g), were treated daily with 2 ml (about 1.67g) undiluted 2-ethylhexanol/kg (applied in drops to the mechanically-shaved dorsal skin, 4 × 4 cm, uncovered) for 12 days (5 days/week over a 16-day period) followed by 14 days of observation. The animals were immobilised for 2 hours after each application. 5 animals were killed on the 17th day (the end of the treatment period), and the remaining 5 on the 30th day (the end of the observation period). On the 10th day a slight reddening and crusting of the skin was evident. The body weights on the 9th and 10th day, and the relative and absolute thymus weights on the 17th day, were significantly ($p \leq 0.05$) reduced. There were no effects on the weights of the heart, liver, spleen or kidneys, the levels of protein, albumin, and α_1-, β_1- and γ-globulin contents in the serum, or the activity of alanine aminotransferase in the serum. Histologically, the following effects were observed (with at least 3 out of 5 treated animals differing from the controls):

Organ findings (Schmidt et al., 1973)

Liver	Histiocytic and imflammatory granulomas, peripheral fine-droplet fatty degeneration
Lungs	Interstitial pneumonia, bronchiectasis, severe round-cell bronchitis
Kidneys	Epithelial-cell necrosis, cysts, basophilic "balloon nuclei"
Heart	Inter- and intracellular oedema, necrobiotic muscle fibres, interstitial oedema
Testes	Interstitial oedema, reduced spermiogenesis
Thymus	Increased "colloidocytes"
Adrenals	Cortex very rich in lipoids

Histochemical investigation of the liver showed raised succinate-dehydrogenase activity and reduced lactate dehydrogenase activity. Tests on acid phosphatase and non-specific α-naphthylacetate esterase activity and on fat coloration gave no indications of any change (Schmidt et al., 1973).

7.3 Skin and mucous membrane effects

Covered application of ca. 0.5 ml undiluted 2-ethylhexanol to the backs of rabbits for one minute had no effect whilst exposure for 5 or 15 minutes resulted in a slight reddening 24 hours after application, which had completely subsided 8 days after exposure. A single 20-hour covered application of approx. 0.5 ml to the backs of rabbits led to a slight reddening of the skin after 24 hours, with oedema formation. 8 days after application, marked flaking was visible at the application site (BASF, 1963).

The administration of 2-ethylhexanol to the uncovered abdominal skin of rabbits caused hyperaemia (no further details; Smyth et al., 1969).

In a recent investigation of skin irritation (OECD method 404), 4-hour exposure to the undiluted product resulted in severe irritation with reddening, oedema and scar-formation (irritation index 6.75, on a scale of 0–8 of increasing irritation (Hüls, 1987).

In an investigation carried out using the Draize method, the instillation of 100 µl 2-ethylhexanol into the conjunctival sac of 6 rabbits led to severe irritation. Corneal clouding, hyperaemia and inflammation of the iris occurred, as well as a conjunctival reddening, oedema formation and mucous secretion (mean irritation values 19 after 24 hours, 20 after 72 hours, and 0 after 7 days; Scala and Burtis, 1973).

In a recent study (OECD method 405), instillation of 0.1 ml undiluted 2-ethylhexanol led to moderately severe symptoms of irritation of the cornea, iris and conjunctiva (irritation index 28.59, on a scale of 0–110 of increasing irritation; Hüls, 1987).

In rabbits, the instillation of 5 µl undiluted 2-ethylhexanol into the conjunctival sac caused slight irritation 24 hours after application, whilst 20 µl caused moderately severe irritation of the cornea (no further details provided in these publications; Smyth et al., 1969; Carpenter and Smyth, 1946).

In a furthter test on the eye of the rabbit, the instillation of 50 µl 2-ethylhexanol caused slight reddening, oedema formation and slight clouding of the cornea. The symptoms had completely subsided 8 days after treatment (BASF, 1963).

The instillation of one drop of a 25 or 50% solution in oil (no further details) or of undiluted 2-ethylhexanol led to reddening and swelling of the conjunctiva, lacrimation and mucous secretion, but there was no damage to the cornea. Recovery occurred within 8 to 96 hours, depending on the concentration. A 12.5% solution caused no signs of irritation (Schmidt et al., 1973).

7.4 Sensitization

No information available.

7.5 Subchronic-chronic toxicity

Rats received 100, 500, 2500 and 12,500 ppm 2-ethylhexanol in their food for 90 days. Mortality, body weight and food uptake were unaffected at all of these concentrations. The animals given the highest dose nevertheless showed signs of liver and kidney damage. The absolute and relative liver weights were increased in both sexes. At dissection, the livers of the female rats were hyperaemic and/or swollen and degenerative changes were evident in the kidneys of the males. Histological examination revealed slight and reversible changes to the liver and kidneys (Union Carbide Corp.).

7.6 Genotoxicity

In vitro. 2-Ethylhexanol was tested on Salmonella typhimurium TA 98, TA 100, TA 1535, TA 1537 and TA 1538 for mutagenic activity (Schimizu et al., 1985; Kirby et al., 1983; Zeiger et al., 1982; CMA, 1982a). In these experiments (plate-incorporation test) concentrations of up to 1500 µg/plate were used. At the high concentrations 2-ethylhexanol had a cytotoxic effect. The experiments were carried out both without and with metabolic activation (S 9 from Aroclor 1254-induced rat and Syrian hamster liver). The studies gave no indication of mutagenic properties.

In a spot test on Salmonella typhimurium strains TA 98, TA 100, TA 1535, TA 1537, TA 1538 and TA 2637, no mutagenic effects occurred with or without metabolic activation (S 9 from Aroclor 1254-induced rat liver). Concentrations of up to 2000 µg/plate were tested. A cytotoxic effect occurred at the high concentrations (Agarwal et al., 1985).

On the other hand, a further study on Salmonella typhimurium TA 100 found very weak mutagenic activity for 2-ethylhexanol in 3 replicate experiments in the absence of a microsomal liver fraction. This effect was dose-dependent at concentrations of 1.0 and 1.5 mM (a µg/plate concentration could not be established, because the publication gives no precise data relating to the conduct of the experiment). The number of back mutants was maximally 3.5 times higher than the values for the blank control. In this experiment, azaguanine resistance (azg-sensitive to azg-resistant) rather than his$^-$ to his$^+$ reversion was measured (Seed, 1982).

In the spot test, 2-ethylhexanol showed no mutagenic potential in B. subtilis H17/M45 at a concentration of 500 µg/plate (Tomita et al., 1982).

No indication of mutagenic activity was obtained in L5178Y mouse lymphoma cells with or without metabolic activation (S 9 from Aroclor 1252/1254-induced rat liver). At the high concentrations (from approx. 0.05 µl/ml=40 µg/ml) cytotoxic effects occurrend (Kirby et al., 1983).

In a chromosomal aberration test in vitro, 2-ethylhexanol showed no activity in CHO-K1-BH4 Chinese hamster cells. The experiment was carried out without metabolic activation. Cytotoxic effects occurred at the higher dosage levels (from approx. 2.2 mM, about 130 µg/ml; Phillips et al., 1982).

In vivo. In the micronucleus test, groups of 6 B6C3F1 mice of each sex received a single i.p. injection of 456 mg 2-ethylhexanol/kg, or the same dose twice with a 24-hour interval. A control group (6 animals/sex) received the solvent maize oil, and a positive control group received triethylene melamine (TEM). Following its administration, 2-ethylhexanol led to apathy of the treated animals, prostration and irregular respiration. The symptoms had subsided after 24 hours. In the males given two doses, there was a significantly ($p \leq 0.05$) increased occurrence of micronuclei in the polychromatic erythrocytes. In the view of the authors, however, this represents a false-positive result due to the excessively low micronuclear values in the control group. The values for the treated groups were within the normal range of the institute (0.28%; CMA, 1982b).

In a cytogenetic test in vivo (chromosomal analysis) on the bone marrow of the rat (Fischer 344), groups of 5 animals received 0.02, 0.07 and 0.2 ml 2-ethylhexanol/kg/day (about 16.7, 58.9 and 167 mg/kg/day, respectively) orally for 5 days. The high dose level corresponded to $^{1}/_{10}$ of the 5-day LD_{50}. 50 metaphases were evaluated per animal. There was no increase in the number of aberrations in the groups treated with 2-ethylhexanol (Putman et al., 1983).

A dominant lethal test with 2-ethylhexanol was carried out with ICR/SIM mice. Male mice received oral doses of 250, 500 or 1000 mg 2-ethylhexanol/kg on 5 concecutive days. The highest level was the maximum tolerated dose. Each treated male was paired with 2 untreated females for 8 cycles, each of 1 week. Dissection of the female mice took place on the 14th to 17th day of gestation. For the groups treated with 2-ethylhexanol, the fertility index and the number of live and dead foetuses lay within the normal range. No treatment-related dominant lethal effects were observed (Rushbrook et al., 1982).

In order to evaluate possible mutagenic properties, the urine of 2-ethylhexanol-treated rats was tested on Salmonella typhimurium.

6 female rats received 1000 mg 2-ethylhexanol/kg/day by oral administration for 15 days. 6 further animals served as controls. The urine of each animal was collected each day and pooled at the end of the experiment. Urine samples treated with the enzymes β-glucuronidase/aryl sulphatase, as well as untreated samples, were tested. All the experiments were carried out with and without metabolic activation. In Salmonella typhimurium strains TA 98, TA 100, TA 1535, TA 1537 and TA 1538, no treatment-related mutagenic activity was evident in comparison with the control, up to a maximum test concentration of 2 ml urine/plate (Divincenzo et al., 1983, 1985).

Mice received food containing 1% (10,000 ppm) diethylhexyl adipate for 4 weeks and then a single dose of 110 or 120 mg ^{14}C-2-ethylhexanol/kg by stomach tube. Only minimal radioactivity, attributable to the incorporation of 2-ethylhexanol metabolites, could be detected in the liver DNA. In rats pretreated for 4 weeks with diethylhexyl phthalate (1% in the food), similarly no activity in the liver attributable to covalent binding to DNA was detectable following single oral administration by stomach tube of 51 or 53 mg ^{14}C-ethylhexanol/kg (v. Däniken et al., 1984).

7.7 Carcinogenicity

2-Ethylhexanol was tested on BALB/3T3 cells (transformed mouse fibroblasts) both without and with metabolic activation (primary rat hepatocytes). At the high concentrations (from 0.093 µl/ml=78 µg/ml without metabolic activation and from approx. 0.048 µl/ml = 40 µg/ml with metabolic activation) there was a cytotoxic effect. Neither study gave any indication of cell transformation by 2-ethylhexanol (CMA, 1982c, 1983).

2-Ethylhexanol was further tested on epidermal-cell cultures of mice (JB6 cells, cell lineage 41). 2-Ethylhexanol showed no transformation potential at concentrations ranging from 41×0^{-7} to 77×10^{-7} mol/l = 5.2×10^{-5} to 1×10^{-3} mg/ml (Ward et al., 1986).

Long-term experiments on the carcinogenicity of 2-ethylhexanol are not so far available, but are in preparation at the request of the EPA. The product is to be administered orally to B6C3F1 mice and Fisher-344 rats, either by stomach tube or in microcapsules in the food (EPA, 1987).

7.8 Reproductive toxicity

In a screening test, mice (Charles River CD-1) received 1525 mg 2-ethylhexanol/kg body weight/day by stomach tube from the 7th to the 14th day of gestation. Mothers and young were observed up to the 3rd day of lactation. 50 animals were used per dosage level and control group. Pregnancy was not checked at the beginning of the experiment. "Timed- pregnant" animals were used. The dosage level used was in the maternally toxic range, the symptoms of toxicity including reduced mobility, ataxia, hypothermia, unkempt appearance and blood in the urine. The body weight of the mothers was significantly reduced. Of the 50 animals treated, 18 died during the experiment, the authors attributing 17 of these deaths to 2-ethylhexanol. None of the control animals died. The fertility index for the treated animals was 40% (control: 68%), the pregnancy index 55% (control 97%, $p \leq 0.05$). There was a significant reduction ($p \leq 0.05$) in the number of living young and in their body weights, in comparison with the control group. No other parameters were investigated in this study (NIOSH, 1983).

Wistar rats received a single dose of 6.25 or 12.5 mmol (about 833 or 1666 mg/kg) by stomach tube on the 12th day of gestation, and were killed on the 20th day. Following caesarian section, the group given 833 mg/kg showed a slight increase of 2% ($p \leq 0.01$) in malformed foetuses relative to the controls (0%). The other parameters (implantation index, mean foetal weight, number of dead and resorbed foetuses) were unaffected. Simultaneous intraperitoneal administration of 150 mg caffeine/kg potentiated this effect (increase in malformed foetuses to $21.2 \pm 10.9\%$). Even after a dose of 1666 mg/kg, the implantation index and percentage of dead and resorbed foetuses were unchanged, although the mean foetal weight, at 3.5 g, was reduced relative to the controls (4.1 g). 22.2 ($\pm$ 14.7)% of the surviving foetuses showed malformations (controls 0%). These included hydronephrosis (7.8%), tail anomalies (4.9%), anomalies of the extremities (9.7%) and "others" (1%). The dose of 1666 mg/kg does, however, correspond to about half the LD_{50} (Ritter et al., 1987).

7.9 Effects on the immune system

No information available.

7.10 Other effects

In vitro studies on rat hepatocytes gave no indications of peroxisome proliferation (marker: palmitoyl-CoA-oxidase). The highest concentration tested was 0.5 mM (about 65.1 µg/ml), the incubation time 48 to 72 hours (Rhodes et al., 1984; Mitchell et al., 1985).

Table 1. Metabolism of 2-ethylhexanol

Species	Rat	Rat[a]	Rabbit	Rabbit	Rabbit
Dose	275 mg/kg	8 mg/animal	3 ml/animal	80 mg/kg	ca 1085 mg/kg
Route of admin.	oral	i.p.	oral	i.p.	oral
Period (hours)	0–18	0–24	0–24	0–24	0–24
Metabolites	% of the metabolites in the urine				
2-ethylhexanol	ca 3	ca 15		0.5	
2-ethylhexanoic acid	ca 60	ca 29		–	
2-ethylhexanoic acid glucuronide		ca 56	ca 90	ca 93	86.9
2-ethyl-2,3-dihydroxy-hexanoic acid				ca 3.5	
2-ethyl-1,6-hexanedionic acid	ca 22–27				
2-ethyl-5-hydroxy-hexanoic acid	ca 12–15				
2-ethyl-5-oxohexanoic acid	ca 1				
2-ethylhexylsulfate				ca 3.0	
2-heptanone	ca 8				
4-heptanone					
Reference	Albro, 1975	Knaak et al., 1966	Kamil et al., 1953b	Knaak et al., 1966	Kamil et al., 1953a

[a] 2-ethylhexanol was tested as the sulphate
– not detected

Table 2. Acute toxicity of 2-ethylhexanol

Species	Number of animals per dose	Sex	Route of administration	Period of subsequent observation	No effect level	LD$_{50}$ (95% confidence)	References
Mouse	$\Sigma^1$70	male	oral[1]	1	—	3830 mg/kg (3280–4460)	Schmidt et al., 1974
Mouse	$\Sigma^1$70	male	oral[1]	2	—	3580 mg/kg (2900–4420)	Schmidt et al., 1974
Mouse	$\Sigma^1$70	female	oral[5]	1	—	2500 mg/kg (2090–3010)	Schmidt et al., 1974
Mouse	$\Sigma^1$50	female	oral[1]	1	—	3220 mg/kg (2870–3610)	Schmidt et al., 1974
Mouse	$\Sigma^1$70	male	i.p.[1]	1	—	891 mg/kg (845–939)	Schmidt et al., 1974
Mouse	$\Sigma^1$50	female	i.p.[1]	1	—	750 mg/kg (726–801)	Schmidt et al., 1974
Mouse	5–30	male/female	i.p.[5]	—	0.42 ml ($\hat{=}$350 mg/kg)	780 mg/kg	Hodge, 1943
Mouse	—	male/female	i.p.[2]	—	—	0.45 ml/kg ($\hat{=}$375 mg/kg)	BASF, 1963
Rat (Sprague-Dawley)	5	male	oral	7–14	—	3750 mg/kg	Scala and Burtis, 1973

Table 2 (continued)

Species	Number of animals per dose	Sex	Route of administration	Period of subsequent observation	No effect level	LD$_{50}$ (95% confidence)	References
Rat (Carworth-Wistar)	5	male	oral	14	–	2.46 ml/kg 2049 mg/kg (1516–2774)	Smyth et al., 1969
Rat	$\Sigma^2$70	female	oral[1]	1	–	4620 mg/kg (3870–5520)	Schmidt et al., 1974
Rat	$\Sigma^2$70	female	oral[1]	2	–	4010 mg/kg (3460–4620)	Schmidt et al., 1974
Rat (Albino)	5–15	male/female	oral	–	0.97 ml/kg ($\hat{=}$808 mg/kg)	3.9 ml/kg ($\hat{=}$3200 mg/kg)	Hodge, 1943
Rat	$\Sigma^2$36	male	oral	14	–	3290 mg/kg (2870–3790)	Schmidt et al., 1973
Rat	–	male/female	oral[3]	7	–	ca. 6.0 ml/kg ($\hat{=}$5000 mg/kg)	BASF, 1963
Rat	$\Sigma^2$50	male	i.p.[4]	1	–	937 mg/kg (860–1020)	Schmidt et al., 1974
Rat	$\Sigma^1$50	female	i.p.	1	–	658 mg/kg (586–739)	Schmidt et al., 1974
Rat (Albino)	5–15	male/female	i.p.	–	0.26 ml/kg ($\hat{=}$217 mg/kg)	0.78 ml/kg ($\hat{=}$650 mg/kg)	Hodge, 1943
Rat	2	male/female	i.p.	14	–	500–1000 mg/kg	Dave and Lidman, 1978

Species	N	Sex	Route	Duration		Value	Reference
Rat	10	male/female	dermal[1] (24 h occlusive)	14	–	>3000 mg/kg	Hüls, 1987
Guinea pig	$\Sigma^3$12	male	oral[1]	1	–	1860 mg/kg (1220–2820)	Schmidt et al., 1974
Guinea pig	–	–	dermal	–	–	>8300 mg/kg	Clayton and Clayton, 1982
Rabbit	$\Sigma^3$12	male	oral[1]	1	–	1180 mg/kg	Schmidt et al., 1974
Rabbit (New Zealand)	4	male	dermal (24 h occlusive)	14	–	2.38 ml/kg (=1980 mg/kg)	Smyth et al., 1969
Rabbit	4	–	dermal (24 h occlusive)	14	–	>2600 mg/kg	Scala and Burtis, 1973

– No further details
Σ^1 50 or 70 animals, no further details
Σ^2 Total of 36 animals, no further details
Σ^3 Total of 12 animals, no further details
1 Undiluted
2 8% aqueous suspension containing traganth
3 30% aqueous suspension containing traganth
4 Diluted 1:5 in arachis oil
5 Diluted 1:10 in maize oil

In another study, there was a 6-fold increase in the carnitine acetyl transferase activity at a concentration of 1 mM (about 130.2 µg/ml; Gray et al., 1983).

Another in vitro study of rat hepatocytes similarly showed significantly (<0.01) increased carnitine acetyl transferase activity at a concentration of 1 mM (about 130.2 µg/ml), as well as increased 7-ethoxy-coumarin-O-deethylase activity and peroxisome proliferation (no further details). The incubation time was 48 hours (Gray et al., 1982).

Incubation of 2-ethylhexanol at concentrations of 2.5 to 15 mM (about 325.6 to 1953.5 µg/ml) with the supernatant liquid of rat liver homogenate (obtained at 9000 g) led to a dose-dependent reduction in aniline hydroxylase and aminopyrine-N-demethylase activity (Agarwal et al., 1982).

In an in vitro model of rat seminiferous tubules, a concentration of 200 µM 2-ethylhexanol (about 26.1 µg/ml) failed to detach sperm from Sertoli cells, in contrast to monoethylhexylphthalate. The incubation time was 48 hours (Gangolli, 1982).

8. Experience in humans

The sensitizing potential of 2-ethylhexanol was tested on 29 subjects (no further details) using the method of Kligman (1966). The concentration used was 4% which, in a preliminary experiment, was found to be slightly irritant (result obtained after 48 hours) and was therefore taken to be the maximum tolerated dose. During the induction phase, a small piece of cloth (surface area ca 9.7 cm^2) soaked in 1.0 ml 5% aqueous sodium laryl sulphate solution was applied to either the forearm or the thigh. This treatment caused slight inflammation. A piece of cloth soaked in 1.0 ml 4% 2-ethylhexanol solution (dissolved in liquid paraffin, USP) was then applied to these subjects at the same site for 28 hours. This two-stage administration procedure was repeated 4 times. The challenge treatment involved 48-hr application, to the skin of the back, of a piece of cloth (surface area ca 6.5 cm^2) soaked in 0.4 ml 4% 2-ethylhexanol. The skin reaction was evaluated after 48 and 96 hours. No sensitization reaction to 2-ethylhexanol was evident in any of the subjects (Opdyke, 1979).

No indications of sensitization have been reported in practice (BASF, 1988).

References

Agarwal, D.K., Agarwal, S., Seth, P.K.
Effect of di(2-ethylhexyl)phthalate on drug metabolism, lipid peroxidation, and sulfhydryl content of rat liver
Drug Metabolism and Disposition, 10, 77–80 (1982)

Agarwal, D.K., Lawrence, W.H., Nunez, L.J., Autian, J.
Mutagenicity evaluation of phthalic acid esters and metabolites in Salmonella typhimurium cultures
J. Toxicol. Environ. Health, 16, 61–69 (1985)

Albro, P.W.
The metabolism of 2-ethylhexanol in rats
Xenobiotica, 10, 625–636 (1975)

BASF AG
Unpublished investigation (1963)

BASF AG
Communication to the Department of Occupational Medicine and Protection of Health (1988)

Carpenter, C.P., Smyth, H.F.
Chemical burns of the rabbit cornea
Am. J. Ophthalmol., 29, 1363–1372 (1946)

Clayton, G.D., Clayton, F.E. (eds.)
Patty's Industrial Hygiene and Toxicology Volume 2C, 4620–4708 (1982)
John Wiley and Sons, New York

CMA
Communication from the Chemical Manufacturers Association to the Employment Accident Insurance Fund of the Chemical Industry. Mutagenicity evaluation of 2-ethyl hexanol (2-EH) in the salmonella/microsome plate test (1982a)

CMA
Communication from the Chemical Manufacturers Association to the Employment Accident Insurance Fund of the Chemical Industry. Mutagenicity evaluation of 2-ethylhexanol (2-EH) in the mouse micronucleus test (1982b)

CMA
Communication from the Chemical Manufacturers Association to the Employment Accident Insurance Fund of the Chemical Industry.

Evaluation of 2-ethylhexanol in the in vitro transformation of BALB/3T3 cells assay (1982c)

CMA
Communication from the Chemical Manufacturers Association to the Employment Accident Insurance Fund of the Chemical Industry.
Evaluation of 2-Ethylhexanol in the in vitro transformation of BALB/3T3 cells with metabolic activation by primary rat hepatocytes (1983)

Cohen, A.J., Grasso, P.
Review of the hepatic response to hypolipidaemic drugs in rodents and assessment of its toxicological significance to man food cosmetic Toxicology, 19, 585–606 (1981)

v. Däniken, A., Lutz, W.K., Jäckh, R., Schlatter, C.
Investigation of the potential for binding of di(2-ethylhexyl)phthalate (DEHP) and di(2-ethylhexyl)adipate (DEHA) to liver DNA in vivo
Toxicol. Appl. Pharmacol., 73, 373–387(1984)

Dave, G., Lidman, U.
Biological and toxicological effects of solvent extraction chemicals range finding acute toxicity in the rainbow trout (Salmo gairdnerii Rich.) and in the rat (Rattus norwegicus L.)
Hydrometallurgy, 3, 210–216 (1978)

Divincenzo, G.D., Hamilton, M.L., Mueller, K.R., Donish, W.H., Barber, E.D.
Bacterial mutagenicity testing of urine from rats dosed with 2-ethylhexanol derived plasticizers
Toxicology, 34, 247–259 (1985)
See also: Divincenzo, G.D., Donish, W.H., Mueller, K.R., Hamilton, M.L., Barber, E.D.
Mutagenicity testing of urine from rats dosed with 2-ethylhexanol derived plasticizers
Environ. Mutagen., 5, 471 (1983)

EPA (U.S. Environmental Protection Agency)
2-Ethylhexanol, Final Test Rule Federal Register, Vol.52, No. 148, dated 3.8.1987

Gangolli, S.D.
Testicular effects of phthalate esters
Environ. Health Perspect., 45, 77–84 (1982)

Ganning, A.E., Klasson, E., Bergman, A., Brunk, U., Dallner, G.
Effect of phthalate ester metabolites on rat liver
Acta Chem. Scand., 36, 563–565 (1982)

Gray, T.J.B., Beamand, J.A., Lake, B.G., Foster, J.R., Gangolli, S.D.
Peroxisome proliferation in cultured rat hepatocytes produced by clofibrate and phthalate ester metabolites
Toxicol. Lett., 10, 273–279 (1982)

Gray, T.J.B., Lake, B.G., Beamand, J.A., Foster, J.R., Gangolli, S.D.
Peroxisomal effects of phthalate esters in primary cultures of rat hepatocytes
Toxicology, 28, 167–179 (1983)

Hammock, B.O., Ota, K.
Differential induction of cytosolic epoxide hydrolase, microsomal epoxide hydrolase, and glutathione S-transferase activities
Toxicol. Appl. Pharmacol., 71, 251–265 (1983)

Hodge, H.C.
Acute toxicity for rats and mice of 2-ethyl hexanol and 2 ethyl hexyl phthalate
Proc. Soc. Exp. Biol. Med., 53, 20–23 (1943)

Hodgson, J.R., Myhr, B.C., McKoen, M., Brusick, D.J.
Evaluation of Di-(2-Ethylhexyl)phthalate and its Major Metabolites in the Primary Rat Hepatocyte Unscheduled DNA Synthesis Assay
Environ. Mutagen., 4, 388 (1982)

Hollenbach, K., Schmidt, P., Stremmel, D.
Tierexperimentelle Untersuchungen zur Blutdruckwirksamkeit von Thioglykolsäureisooctylester, Thioglykolsäure und 2-Äthylhexanol
Z. Gesamte Hyg., 18, 481–485 (1972)

Hüls AG
Unpublished report (1987)

Kamil, I.A., Smith, J.N., Williams, R.T.
Studies in detoxication
47. The formation of ester glucuronides of aliphatic acids during the metabolism of 2-ethylbutanol and 2-ethylhexanol
Biochem. J., 53, 137–140 (1953a)

Kamil, I.A., Smith, J.N., Williams, R.T.
46. The metabolism of aliphatic alcohols the glucuronic acid conju-

gation of acyclic aliphatic alcohols
Biochem. J., 53, 129–136 (1953b)

Keith, Y., Canning, P.M., Lhuguenot, J.-C., Elcombe, C.R.
Peroxisome proliferation due to di(2-ethylhexyl)adipate and 2-ethylhexanol
Hum. Toxicol., 4, 551–552 (1985)

Kirby, P.E., Pizzarello, R.F., Lawlor, T.E., Haworth, S.R., Hodgson, J.F., Mason, G.
Evaluation of di(2-ethylhexyl)phthalate and its major metabolites in the ames test and L5178Y mouse lymphoma mutagenicity assay
Environ. Mutagen.,4, 388–389 (1982)

Kirby, P.E., Pizzarello, R.F., Lawlor, T.E., Haworth, S.R., Hodgson, J.R.
Evaluation of di(2-ethylhexyl)phthalate and its major metabolites in the ames test and L5178Y mouse lymphoma mutagenicity assay
Environ. Mutagen., 5, 657–663 (1983)

Kligman, A.M.
The identification of contact allergens by human assay III. The maximization test: A procedure for screening and rating contact sensitizer
J. Invest. Dermatol., 47, 393–409 (1966)

Knaak, J.B., Kozbelt, S.J., Sullivan, L.J.
Metabolism of 2-ethylhexyl sulfate by the rat and rabbit
Toxicol. Appl. Pharmacol., 8, 369–379 (1966)

Lake, B.G., Gangolli, S.D., Grasso, P., Lloyd, A.G.
Studies on the hepatic effects of orally administered di(2-ethylhexyl)-phthalate in the rat
Toxicol. Appl. Pharmacol., 32, 355–367 (1975)
See also: Lake, B.G., Gangolli, S.D., Wright, M.G., Grasso, P., Lloyd, A.G.
Biochem. Soc. Trans. 2, 322–325 (1974)

Lillard, D.A., Powers, J.J.
Aqueous odour threshold of organic pollutants in industrial effluents
National Environmental Research Center,
Report PB-242 734 (1975)

Merck Index, The
10th Edition Merck & Co, Inc., Rahway (1983)

Mitchell, A.M., Lhuguenot, J.-C., Bridges, J.W., Elcombe, C.R.
Identification of the proximate peroxisome proliferator(s) derived from di(2-ethylhexyl)phthalate
Toxicol. Appl. Pharmacol., 80, 23–32 (1985)

Moody, D.E., Reddy, J.K.
Hepatic peroxisome (microbody) proliferation in rats fed plasticizers and related compounds
Toxicol. Appl. Pharmacol., 45, 497–504 (1978)
See also: Moody, D.E., Azarnoff, D.L., Reddy, J.K.
Induction of hepatic peroxisomes, peroxisome associated enzymes and hypolipidemia in rats treated with 2-ethyl hexanoic acid and 2-ethyl hexanol
J. Cell. Biol., No. 1088 (1976)

Moody, D.E., Reddy, J.K.
Serum triglyceride and cholesterol contents in male rats receiving diets containing plasticizers and analogues of the ester 2-ethylhexanol
Toxicol. Lett., 10, 379–383 (1982)

Muller, J., Greff, G.
Recherche du relation entre toxicité de molécules d'interêt industriel et propietés physico-chimiques: test d'irritation des voies aeriennes superieurs applique à quatre familles chimiques
Food Chem. Toxicol., 22, 661–668 (1984)

NAPIRI (National Association of Printing Ink Research Institute)
Raw Material Data handbook-Organic Solvents
National Association of Printing Ink Research Institute Data sheet 1–61 (1974)

NIOSH (National Institute for Occupational Safety and Health)
Screening of priority chemicals for potential reproductive hazard
Hazleton Study No. 6125–101 through 6125–110, Report PB 85–220143 (1983)

Opdyke, D.L.J.
Fragrance Raw Materials Monographs 2-Ethylhexanol
Food Cosm. Toxicol., 17, 775–777 (1979)

Phillips, J.B., James, T.E.B., Gangolli, S.D.
Genotoxicity studies of di(2-ethylhexyl)phthalate and its metabolites in CHO Cells
Mutat. Res., 102, 297–304 (1982)

Putman, D.L., Moore, W.A., Schechtmann, L.M., Hodgson, J.R.
Cytogenetic evaluation of di(2-ethylhexyl)phthalate and its major metabolites in Fischer 344 rats
Environ. Mutagen., 5, 227–231 (1983)

Rhodes, C., Soames, T., Stonard, M.D., Simpson, M.G., Vernall, A.J., Elcombe, C.R.
The absence of testicular atrophy and in vivo and in vitro effects on hepatocyte morphology and peroxisomal enzyme activities in male rats following the administration of several alcohols
Toxicol. Lett., 21, 103–109 (1984)

Ritter, E.J., Scott, W.J.Jr., Rondall, J.L., Ritter, J.M.
Teratogenicity of di(2-ethylhexyl)phthalate, 2-ethylhexanol, 2-ethylhexanoic acid and valproic acid and potentiation by caffeine
Teratology, 35, 41–46 (1987)

Rushbrook, C.J., Jorgenson, T.A., Hodgson, J.R.
Dominant lethal study of di(2-ethylhexyl)phthalate and its major metabolites in ICR/SIM mice
Environ. Mutagen., 4, 387 (1982)

Scala, R.A., Burtis, E.G.
Acute toxicity of a homologous series of branched-chain primary alcohols
Am. Ind. Hyg. Assoc. J., 34, 493–499 (1973)

Schmidt, P., Gohlke, R., Rothe, R.
Zur Toxizität einiger C8-Aldehyde und Alkohole
Z. Gesamt Hyg., 19, 485–490 (1973)

Schmidt, R., Fox, G., Hollenbach, W., Roth, R.
Zur akuten Toxizität des Thioglykolsäure-2-äthylhexylesters im Tierversuch
Z. Gesamt Hyg., 20, 575–578 (1974)

Seed, J.L.
Mutagenic activity of phthalate esters in bacterial liquid suspension assay
Environ. Health Perspect., 45, 111–114 (1982)

Shimizu, H., Suzuki, Y., Takemura, N., Goto, S., Matsushita, H.
The results of microbial mutation test for forty-three industrial chemicals
Jpn. J. Ind. Health, 27, 400–419 (1985)

Sjöberg, P., Bondesson, U., Gray, T.J.B., Plöen, L.
Effects of di(2-ethylhexyl)phthalate and five of its metabolites in rat testis in vivo and in vitro
Acta Pharmacol. Toxicol., 58, 225–233 (1986)

Smyth, H.F., Carpenter, C.P., Weil, C.S., Pozzani, U.C., Striegel, J.A., Nycum, J.S.
Range-Finding Toxicity Data: List VII
Am. Ind. Hyg. Assoc. J., 30, 470–476 (1969)

Tomita, I., Nakamura, Y., Aoki, N., Inui, N.
Mutagenic/carcinogenic potential of DEHP and MEHP
Environ. Health Perspect., 45, 119–125 (1982)

Ullmann's Encyclopedia of Industrial Chemistry
4th Edition, Volume 16, 304–309 (1975)
Verlag Chemie, Weinheim

Ullmann's Encyclopedia of Industrial Chemistry
5th Edition, A10, 137–141 (1987)
VCH Verlagsgesellschaft mbH, Weinheim

Union Carbide Corporation
Unpublished data cited in Clayton and Clayton, 1982

Ward, J.M., Diwan, B.A., Ohshima, M., Hu, H., Schuller, H.M., Rice, J.M.
Tumor-initiating and promoting activities of di(2-ethylhexyl)phthalate in vivo and in vitro
Environ. Health Perspect., 65, 279–291 (1986)

Zeiger, E., Haworth, S., Speck, W., Mortelmans, K.
Phthalate ester testing in the national toxicology program's environmental mutagenesis test development program
Environ. Health Perspect., 45, 99–101 (1982)
See also: Zeiger, E., Haworth, S., Mortelmans, K., Speck, W.
Mutagenicity testing of di(2-ethylhexyl)phthalate and related chemicals in Salmonella
Environ. Mutagen., 7, 213–232 (1985)

Zoeteman, B.C.J., Piet, G.J., Rygrok, C.T.M., Van de Heuvel, R.
Threshold odour concentrations in water of chemical substance annex to Bulletin No 73–7
National Institute for Water Supply, The Hague, The Netherlands (1974), cited in Lillard and Powers (1975)

Butynediol

1. Summary and assessment

According to the available findings on acute toxicity, butynediol is toxic on oral administration (LD_{50} in the rat orally 105–135 mg/kg). It causes central nervous system depression and in higher doses also excitation.

Depending on its concentration, butynediol can produce definite skin irritation. However, its irritative effect on the eye is slight.

With longer-term oral administration of 0.04 and 0.2 mg/kg body weight/day over 6 months, no characteristic toxic effects are observed in rats. A concentration of 2 mg/kg body weight leads to impairment of the conditioned reflexes, enzyme changes in the serum and histological changes in the brain and liver.

Butynediol is not mutagenic in the Salmonella/microsome test. A chromosome aberration test in vitro with V79 cells of the chinese hamster has not shown any clastogenic effect.

One case of human sensitization to butynediol has been observed. A sensitization experiment with guinea pigs is currently being conducted on behalf of the BG Chemie.

2. Name of substance

2.1 Usual name   Butynediol
2.2 IUPAC-name   2-Butyne-1,4-diol
2.3 CAS-No.   110-65-6

3. Synonyms

Common and trade names   1,4-dioxy-butyne-2
                          bis-oxy methylacetylene
                          but-2-yne-1,4-diol
                          Golpanol
                          Korantin BH

4. Structural and molecular formulae

4.1 Structural formula

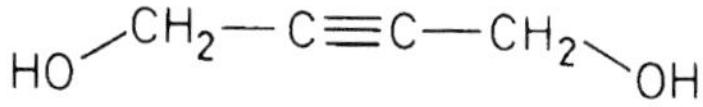

HO–CH$_2$–C≡C–CH$_2$–OH

4.2 Molecular formula

C$_4$H$_6$O$_2$

5. Physical and chemical properties

5.1	Molecular mass	86.09 g/mol
5.2	Melting point	57.5° C
5.3	Boiling point	238° C
5.4	Density	1.04–1.05 g/cm^3 (at 20° C)
5.5	Vapour pressure	2.3 hPa (at 120° C)
5.6	Solubility in water	374 g/100 g water (at 25° C)
5.7	Solubility in organic solvents	readily soluble in alcohols and acetone hardly soluble in ether and hydrocarbons almost insoluble in benzene
5.8	Solubility in fat	no information available
5.9	pH-value	approx 3 (at 100 g/liter water)
5.10	Conversion factors	1 ppm $\hat{=}$ 3.578 mg/m^3 1 mg/m^3 $\hat{=}$ 0.280 ppm

Ullmann, 1975; BASF, 1987; Sax, 1979; Knyshova, 1968

6. Uses

As a glazing agent in electroplating, corrosion inhibitor, stabilizer for halogenated hydrocarbons; intermediate product in the manufacture of 2-butene-1,4-diol, 1,4-butanediol, insecticides, herbicides (Ullmann, 1985).

7. Results of experiments

7.1 Toxicokinetics and metabolism

Butynediol is oxidised in the rat by alkohol dehydrogenase (E.C.1.1.1.1) to breakdown products that have not so far been further identified. The Michaelis constant (Km) for this reaction (under in vitro conditions) was 8.2×10^{-4} M (for comparison, ethanol 7.9×10^{-4} M) (Taberner and Pearce, 1974).

7.2 Acute and subacute toxicity

The LD_{50} for mice was about 105 mg/kg body weight following oral dosing and about 100 mg/kg body weight following intraperitoneal injection, for the rat between 100 and 135 mg/kg body weight following oral dosing and between 52 and 55 mg/kg body weight following dosing intraperitoneally, and for the rabbit about 150 mg/kg body weight following oral dosing.

See table for relevant details.

In a preliminary study, a single oral administration of 50 mg/kg body weight was lethal to cats, whilst rabbits survived the same dose given once or twice. In the case of the rabbit, a single administration of 100 mg/kg body weight led to the death of the animals (no further details, BASF, 1986).

Acute poisoning symptoms included sedation, analgesia, balance disturbances, lying on the side, tonic-clonic convulsions, ruffled coat, accelerated respiration, bradycardia, apathy, salivation and, in the case of rats, coughing (according to the authors) and diarrhoea (Taberner and Pearce, 1974; BASF, 1986).

On macroscopic examination, isolated haemorrhages in the gastrointestinal tract, fatty degeneration of the liver and pulmonary oedema were evident (BASF, 1986).

Butynediol leads to marked hypothermia in rats. An intraperitoneal injection of 0.408 mmol butynediol/kg body weight ($\hat{=}$ 35 mg/kg body weight) caused a reduction in body temperature to approx 35° C 2.5 hours after dosing, whilst a dose of 1.634 mmol/kg body weight intraperitoneal (i.p.) ($\hat{=}$ 141 mg/kg body weight) caused the body temperature to fall to 30.1° C. In the animals receiving the low dose the body temperature returned to normal within 12 hours. The animals receiving the high dose died within 2.5 to 3 hours after administration (Taberner and Pearce, 1974).

In the inhalation risk test (20° C) all 12 rats survived the 8-hour exposure without symptoms (BASF, 1986; Ullmann, 1985). The lowest published lethal concentration (LCLo) for 2-hour inhalation of butynediol was given as 150 mg/m^3 ($\hat{=}$ 42 ppm) for the mouse and the rat (no further details, Ismerov, 1982).

Table 1. Acute toxicity of butynediol

Animal species	Number animals/dose	Sex	Route of administration	Subsequent observation period	No effect level	LD_{50}/LC_{50}	References
Mouse	–	–	oral	–	–	104.75 mg/kg bw	Knyshova, 1968
Mouse	–	–	i.p.	7 days	–	ca 100 mg/kg bw	BASF, 1986
Mouse	–	–	inhalation (2 hr)	–	–	150 mg/m^3 (LCLo)	Ismerov, 1982
Rat	–	–	oral	–	–	104.50 mg/kg bw	Knyshova, 1968
Rat	–	–	oral	7 days	–	0.13 ml/kg bw ($\hat{=}$ 135 mg/kg bw)	BASF, 1986
Rat	–	–	oral	7 days	–	ca 100 mg/kg bw	Ullmann, 1975, 1985; BASF, 1986

Species			Route	Duration			Reference
Rat	—	—	inhalation (2 hr)	—	—	150 mg/m³ (LCLo)	Ismerov, 1982
Rat	6	—	i.p.	24 hr	0.20 mmol/kg bw (≙ 17.2 mg/kg bw)	0.609–0.635 mmol/kg bw (≙ 52.4–54.7 mg/kg bw)	Taberner and Pearce, 1974
Guinea pig	—	—	oral	—	—	130 mg/kg bw	Knyshova, 1968
Rabibt	—	—	oral	—	—	150 mg/kg bw	Knyshova, 1968

7.3 Skin and mucous membrane effects

Undiluted butynediol has a skin-irritating effect (Ullmann, 1975, 1985). Following 4-hour semi-occlusive exposure on the dorsal skin of rabbits, reddening occurred, with some haemorrhaging and moderate to severe oedema. A 30% butynediol solution (in water) was not irritating to rabbit skin (duration of application up to 20 hours, occlusive). On the rabbit eye the irritant effect of the undiluted product was only slight. There was merely a slight reddening of the conjunctiva. The 30% aqueous solution was not irritating (BASF, 1986).

7.4 Sensitization

No information available.

7.5 Subchronic/chronic toxicity

In a study, oral administration of butynediol to male rats (6 animals/dosage level) at doses of 0, 0.04, 0.2 and 2 mg/kg body weight for 6 months did not alter general behaviour, body weight or blood values (haemoglobin content, numbers of erythrocytes, leucocytes and thrombocytes, and coagulation time). At 2 mg/kg body weight the conditioned reflexes were delayed, with a 40% increase in the latency period. In addition, reduced cholinesterase and SH-enzyme activities and increased transaminase activities were seen, as well as an alteration in the serum protein profile. In the brain, the high dosage level led to a reduced number of Nissl's bodies and an increase in the neuroglia, as well as reduced SH-enzyme activity.

In the liver, there was fatty degeneration, sclerotic zones and reduced glycogen values, while in other organs localized hyperaemia occurred (no further details, Knyshova, 1968).

7.6 Genotoxicity

In vitro. Butynediol (purity level >99%) was tested on the *Salmonella typhimurium* strains TA 1535, TA 1537, TA 1538, TA 98 and TA 100 for mutagenic activity (concentration range 20 to 5000 µg/plate, with and without metabolic activation). Butynediol exhibited no mutagenic effect (BASF, 1986).

Butynediol was assessed for its potential to induce structural chromosome aberrations in V79 cells of the Chinese hamster in vitro. Preparation of chromosomes was done 7 h (high dose), 18 h (low, medium and high dose), and 28 h (high dose) after start of treatment with the test article in two (without S9 mix) and three (with S9 mix) independent experiments. The treatment interval was 4 h. In each experimental group two parallel cultures per experiment were used. Per culture 100 metaphases were scored for structural chromosomal aberrations.

The following dose levels were evaluated:

Experiment I and *II*
    without S9 mix:
        7 h: 860.0 µg/ml
        18 h: 50.0; 300.0; 860.0 µg/ml
        28 h: 860.0 µg/ml

Experiment I, II and *III*
    with S9 mix:
        7 h: 300.0 µg/ml
        18 h: 10.0; 100.0; 300.0 µg/ml
        28 h: 300.0 µg/ml

The concentration range of the test article applied had been determined in a pre-experiment using the plating efficiency assay as indicator for toxicity response. Treatment with 860 µg/ml and 300 µg/ml, respectively, in the absence and presence of S9 mix reduced clearly the plating efficiency of the V79 cells. Also the mitotic index was reduced after treatment with these concentrations. There was no relevant reproducible increase in cells with structural chromosome aberrations after treatment with Butynediol. Appropriate reference mutagens were used as positive controls and showed distinct increases in cells with structural chromosome aberrations. In conclusion, it can be stated that in the study described and under the experimental conditions reported, the test article did not induce structural chromosome aberrations as determined by the chromosomal aberration test in the V79 Chinese hamster cell line (Heidemann, 1989).

In vivo.
No information available.

7.7 Carcinogenicity

No information available.

7.8 Reproductive toxicity

No information available.

7.9 Other effects

No information available.

8. Experience in humans

The odour threshold for humans was given as 200 mg butynediol/liter water (no further details, Knyshova, 1968).

A 41-year-old woman developed a pruritic eczema on the face, hands and forearms through contact with butynediol as a component of a cleansing agent. The symptoms appeared approximately 12 hours after contact. Standard patch tests were negative. However, the test with a 10% solution of the cleansing agent was positive. In tests with the different components of this agent, only butynediol (0.01% in water) proved positive, the degree of purity of the butynediol tested being 99.9% (Baadsgaard and Jörgensen, 1985).

References

Baadsgaard, U., Jörgensen, J.
Contact dermatitis to butyne-2-diol-1,4
Contact Dermatitis, 13, 34 (1985)

BASF AG
Unpublished studies (1973, 1981, 1986)

BASF AG
DIN-Safety data sheet "Butin-2-diol-1,4", 05/87

Heidemann, A.
Chromosome aberration assay in chinese hamster V79 cells in vitro with Butynediol
Cytotest Cell Research, Roßdorf, project 137406
on behalf of BG Chemie

Ismerov, N.F.
Toxicometric parameters of industrial toxic chemicals under single exposure
Moscow (1982)
cited in RTECS, 1987

Knyshova, S.P.
Biological effect and hygienic significance of 1,4-butynediol and 1,4-butanediol
Hygiene and Sanitation, 33, 41–47 (1968)

RTECS (Registry of Toxic Effects of Chemical Substances)
U.S. National Institute for Occupational Safety and Health, Cincinnati, Ohio (1986), Suppl. 1987

Sax, N.J. (ed.)
Dangerous Properties of Industrial Materials, 5th edition
van Nostrand Reinhold Company (1979)

Taberner, P.V., Pearce, M.J.
Hypothermic and toxic actions of 2-butyne-1,4-diol and other related diols in the rat
J. Pharm. Pharmac., 26, 597–604 (1974)

Ullmanns Enzyklopädie der technischen Chemie
4th Edition, vol. 9, 19–24 (1975)
Verlag Chemie, Weinheim

Ullmann's Encyclopaedia of Industrial Chemistry
5th edition, vol. A4, 455–462 (1985)
VCH Verlagsgesellschaft, Weinheim

Diethylene glycol

1. Summary and assessment

Between 40 and 70% of an administered dose of diethylene glycol (DEG) is eliminated unchanged by the kidneys. The half-life in the blood of the rat is between 8 and 12 hours, depending on the dose. Single and repeated applications of DEG to the rabbit and dog do not increase the oxalic acid concentration in the urine, whereas in rats under similar experimental conditions there is a dose-dependent increase in urinary oxalic acid. Approx. 0.1% or less of the administered DEG is recovered in the urine of rats in the form of oxalic acid. Metabolic acidosis is observed in the animals. There are no further investigations on the metabolism of DEG.

DEG is of low acute toxicity in the mouse, rat, guinea pig, rabbit, dog and cat (oral LD_{50} rat, approx 15 ml/kg). Narcosis is the most obvious effect of acute poisoning resulting from high doses of DEG; lethally dosed animals show progressive loss of consciousness and usually die within a short time (1–2 days). Autopsy findings show degenerative changes in the kidneys and liver. The available data indicate no local irritative effect on the skin or mucosa. DEG can be absorbed through the skin only at high doses and following repeated application.

There is no information on the sensitizing potential of this substance. On repeated administration to the rat, rabbit and dog of doses which do not lead to a narcotic death, the kidney is the main target organ for the toxic effects of DEG. Diuresis is followed by anuria, protein- and haemoglobinuria, loss of consciousness, dyspnoea and coma. Histopathological findings include nephrosis, tubular necrosis, cystitis and bladder stones; there is fatty or vacuolar degeneration of the liver cells. In the mouse, effects on the central nervous system (oedema, hyperaemia and destruction of neurones) have been described.

DEG has shown no indication of mutagenic activity in a range of in vitro assays (Ames test with or without S9-mix, HGPRT test, SCE test, chromosome aberrations in CHO cells).

Oral administration of DEG in the feed (2%, 4%) to male rats over two years leads to extensive damage to the urinary tract system (including nephrosis, tubular necrosis, cylinder formation and cystitis); bladder stones (of calcium oxalate) and bladder tumours (predominantly papillomas and a metastasizing carcinoma) have been found. A further 2-year feeding study (2%, 4%) found the incidence of bladder stones (of calcium oxalate) to be dose-related and, in the 4% group, a microscopically-confirmed bladder tumour was seen.

Correlated with these observations is the increased elimination of calcium oxalate in the urine of rats given high dietary levels of DEG (40,000 ppm in the feed) for 28 days, and the formation of calcium oxalate crystals in the male rats of this group.

The implantation of calcium oxalate stones or glass balls in the bladder, or the undertaking of a sham operation (surgical slitting and ligature of the bladder) leads to the appearance of bladder stones and tumours, whether or not there is additional treatment with DEG. With one exception, all the rats with bladder tumours also had bladder stones.

These findings support the theses of Grasso (1976) and Mastromatteo (1981), as well as the opinion of Altmann et al. (1986), that chronic irritation by a foreign body in the bladder of the rat can lead to the formation of a tumour. On the other hand, Hueper as recently as 1963 still supported the view that bladder stones, irrespective of their composition, were not the primary cause of tumours of the bladder.

Following inhalation of DEG over 7 months and follow-up observations over 2.5–11 months (concentrations 4–5 mg/m^3, approx 0.92 ppm), adenocarcinoma of the breast is reported in seven out of sixteen female mice. This experiment cannot be evaluated due to lack of sufficient data.

Following gavage administration of 0.84 g/kg (3 times a week, 243 times in total) or subcutaneous injection of 2.5 g/kg (once a week, 82 times) to male rats, and following gavage administration of 0.42 g/kg (3 times a week, 198 times) or subcutaneous injection of 1.25 g/kg (once a week) to mice, an increase in the total number of tumours and in haemoblastoses is reported in both rats and mice with either method of administration, in comparison with the controls. Again, there is a lack of important data which are necessary for assessing the results.

Bearing in mind the validity and meaningfulness of all the available findings on carcinogenic and mutagenic effects, it can be said that, in all probability, no primary carcinogenic effect can be ascribed to DEG. There is no indication in the literature of reproductive toxicity, embryotoxicity or teratogenicity in rats treated with DEG. An embryotoxicity/teratogenicity study involving oral administration to rabbits is currently being conducted on behalf of the BG Chemie.

Experience with humans shows that oral DEG in quantities of approx 1 ml/kg (approx 1.12 g/kg) body weight (approx 60–70 ml/person (67–78 g/person)) can lead to severe intoxication with death in some cases (kidney failure due to tubular nephrosis). DEG poisoning in humans is characterized by nausea, dizziness, pain in the kidney region, followed by suppression of urine and death in uraemic coma; depression of the central nervous system is conspicuous following a single high dose. As in animal experiments, severe pathological changes occur in the kidneys and liver. Oxaluria has been reported in humans following high oral doses of DEG. A harmless daily dose for humans is given as not more than 30–60 mg.

There are no indications that bladder stones or tumours occur in humans following exposure to DEG.

2. Name of substance

2.1 Usual name   Diethylene glycol
2.2 IUPAC-name   3-Oxapentane-1,5-diol
2.3 CAS-No.   111-46-6

3. Synonyms

Common and trade names   Bis(2-hydroxyethyl)ether
Brecolane NDG,
Deactivator E
Deactivator H
DEG
DICOL
Diglycol
beta,beta'-Dihydroxydiethyl ether
Dissolvand APU,
Ethanol, 2,2'-oxydi-
Ethylene diglycol
Glycol ether
Glycol ethyl ether
3-Oxa-1,5-pentanediol

Diethylene glycol

Common and trade names   2,2'-Oxybisethanol
2,2'-Oxydiethanol
TL4N

4. Structural and molecular formulae

4.1 Structural formula   HO–CH$_2$–CH$_2$–O–CH$_2$–CH$_2$–OH

4.2 Molecular formula   C$_4$H$_{10}$O$_3$

5. Physical and chemical properties

5.1	Molecular mass	106.12 g/mol
5.2	Melting point	–10.5° C
5.3	Boiling point	245° C
5.4	Density	1.119 g/cm^3 (at 20° C)
5.5	Vapour pressure	approx 0.027 hPa (at 20° C)
5.6	Solubility in water	good
5.7	Solubility in organic solvents	soluble in ethanol and ether solvents
5.8	Solubility in fat	no information available
5.9	pH-value	no information available
5.10	Conversion factors	1 ppm $\hat{=}$ 4.403 mg/m^3 1 mg/m^3 $\hat{=}$ 0.227 ppm (CRC, 1981/82; Ullmann, 1974)

6. Uses

De-icer for take-off and landing runways and aircrafts; solvent; additive in printing inks; for industrial drying of gases; auxiliary in paper and textile manufacture (Ullmann, 1974).

7. Results of experiments

7.1 Toxicokinetics and metabolism
    Winek et al.(1978) gave groups of 12 Sprague-Dawley rats (200–300 g)a single oral dose of 6 ml and 12 ml DEG/kg. The DEG

content in the blood was determined in 2 animals in each case after 1, 4, 8, 12, 24 and 48 hours for both dose groups. The half-life of the test substance was 8 hours for the 6 ml/kg dose and 12 hours for the 12 ml/kg dose. The maximum concentration of oxalate (7 and 6 µg/ml) in the blood was measured 8 hours after administration, for both doses. The maximum concentration of oxalate in the kidneys occurrred in each case 48 hours after administering the substance (12.5 and 14.0 µg/g). After administering 15 ml/kg by stomach tube to 30 male Wistar rats, 66% of the animals died within 5 days; metabolic acidosis occurred. 5.93 mg oxalic acid (meanvalue) was excreted in 55.2 ml urine (mean value) in the first 24 hours, and 2.36 mg oxalic acid (mean value) was excreted in 18.5 ml urine (mean value) in 24–48 hours (Hébert et al., 1978; Durand et al., 1976). Haag and Ambrose (1937) observed a dose-dependent increase of oxalic acid in the urine on each of 3 or 4 rats to which 0.03, 0.3 and 1.0% solutions of DEG had been administered as drinking water for 20 days (total oxalic acid in the urine was 13.02 mg, 13.02 mg and 20.16 mg respectively, control 6.93 mg). Hanzlik et al. (1939) determined the DEG content of the blood of 5 dogs which had received 1.5 ml DEG/kg orally, 3 times daily on 2 consecutive days. 24 hours after the first dose they found 44 to 218 mg/100 ml, after 48 hours 108 to 327 mg/100 ml and after 72 hours 96 to 273 mg/100 ml. The large fluctuations were attributed to the animals' vomiting. 4/5 dogs died after 72 hours. The surviving animal only exhibited values of 39 and 34 mg/100 ml. For elimination of DEG from the blood after a single oral dose of 5 ml DEG/kg, a time of >36 hours was reported for 2 dogs (without further details). Oral administration of 2 ml DEG/kg to 2 dogs did not lead to increased oxalic acid in the animals' urine. 40–70% of the substance administered was excreted unchanged in the urine, and traces were also detected in the faeces. Furthermore, 2 ml DEG/kg given orally, 5 times in one week, did not lead to a definite increase of oxalic acid in the urine (no further details are given, Haag and Ambrose, 1937). Wiley et al. (1938) did not detect an increased oxalic acid content in the urine of 1 dog which had been injected daily with 5 ml DEG for a period of 7 days (no further details). Moreover, after treatment for 7 days with DEG (2 ml daily, route of administration not stated), 2 rabbits showed no increase in oxalic acid in the urine, compared with an untreated control. DEG is excreted by rats and rabbits via the kidneys; the amounts of DEG that appear unchanged in the urine are not stated. Less than 0.1% of the administered dose of DEG was found as oxalic acid in rat urine, and it was concluded from this that DEG is only metabolized to a slight extent (Winek et al., 1978; Remmer, 1985).

p-Dioxan-2-one was identified as a metabolite of DEG. Administration of DEG to male rats at a dose of 150 mg/100 g body weight led to excretion of 38 mg p-dioxan-2-one/100 g body weight. p-Dioxan-2-one was excreted almost completely in the first 16 hours after administration. On the basis of this investigation, the assumption was made that DEG could be a metabolite in the metabolism of p-dioxane (Woo et al., 1977a, b). When DEG, dissolved in demineralized water, was administered by stomach tube to 6 female Sprague-Dawley rats at doses of 0.2, 0.685 and 2.0 g/kg, a significant, dose-dependent increase of lactate dehydrogenase was observed in the 24-hour urine at the top two doses. With the highest dose, urine volume was significantly increased (Freundt et al., 1987). The literature does not contain other investigations relating to the metabolism of DEG. There are no investigations into whether DEG, like ethylene glycol, can be metabolized inter alia to CO_2 (Remmer, 1985; Gessner et al., 1961).

7.2 Acute and subacute toxicity

The LD_{50} values for diethylene glycol (DEG) in various animal species and with different routes of administration are presented in Table 1.

The symptoms of acute DEG intoxication were largely similar for mice, rats, guinea pigs, rabbits and dogs. It led to thirst, diuresis, refusal to eat, then to decreased urine volume, proteinuria, dyspnoea, stupor, weakness, disturbances of coordination, reduced temperature, coma, and death through respiratory or cardiac arrest. Kidney lesions (necrosis of the tubules) and liver lesions (necrosis of the central lobules) were found macroscopically and histologically (Laug et al., 1939; Loeser, 1954; Bornmann, 1954a, b, 1955; Geiling et al., 1937; Wegener, 1953). On single administration of DEG to male Wistar rats (15 ml/kg) and Sprague-Dawley rats (6 and 12 ml/kg), by stomach tube, acute intoxication occurred with metabolic acidoses, leading to tubular necrosis in the kidneys and to precipitation of calcium oxalate crystals in the tubules (Hébert et al., 1978; Durand et al., 1976; Winek et al., 1978). For testing the histopathological changes produced by DEG, 130 rats (breed not stated, ratio of female to male 4:1) were given undiluted DEG by stomach tube in doses of 25.0–7.5 ml/kg for a maximum of 3 weeks; intermediate dissections were performed each week. At doses of 25 and 10 ml/kg,

Table 1. Acute toxicity of diethylene glycol

Species	Route of administration	LD_{50}	Lethal dose	References
Mouse	oral	23.70 mg/kg	—	Laug et al., 1939
Mouse	oral	25.23 ml/kg	—	Loeser, 1954
Mouse	oral	13.30 ml/kg	—	Rowe and Wolf, 1982
Mouse	s.c.	20.16 ml/kg	—	Loeser, 1954
Mouse	s.c.	—	50 ml/kg	Browning, 1965
Mouse	i.p.	9.60 g/kg	—	Karel et al., 1947
Rat	oral	14.80 ml/kg	—	Laug et al., 1939
Rat	oral	25.70 ml/kg	—	Loeser, 1954
Rat	oral	20.76 g/kg	—	Smyth et al., 1941
Rat	oral	15.60 g/kg	—	Rowe and Wolf, 1982
Rat	oral	—	15.0 ml/kg[a]	Haag and Ambrose, 1937
Rat	oral	27.0 ml/kg	—	Weatherby and Williams, 1939
Rat	s.c.	18.06 ml/kg	—	Loeser, 1954
Rat	s.c.	—	5.0 ml/kg[a]	Haag and Ambrose, 1937
Rat	i.m.	—	7.0 ml/kg[a]	Haag and Ambrose, 1937
Rat	i.v.	—	5.0 ml/kg[a]	Haag and Ambrose, 1937
Rat	i.p.	—	6.86 ml/kg	Rowe and Wolf, 1982
Guinea pig	oral	7.76 ml/kg	—	Laug et al., 1939
Guinea pig	oral	13.21 g/kg	—	Smyth et al., 1941
Guinea pig	oral	14.00 g/kg	—	Rowe and Wolf, 1982

Table 1 (continued)

Species	Route of administration	LD$_{50}$	Lethal dose	References
Rabbit	oral	4.40 ml/kg	–	Laug et al., 1939
Rabbit	i.v.	1.0–2.0 ml/kg	–	Browning, 1965; Kesten et al., 1937
Rabbit	i.v.	–	2.0 ml/kg[a]	Haag and Ambrose, 1937
Rabbit	i.m.	–	4.0 ml/kg[a]	Haag and Ambrose, 1937
Rabbit	dermal	11.9 ml/kg	–	Rowe and Wolf, 1982
Dog	oral	10.0 ml/kg	–	Laug et al., 1939
Dog	i.v.	5.8 ml/kg	–	Weatherby and Williams, 1939
Cat	oral	–	ca 3.3–4.7 ml/kg	Laug et al., 1939

[a] Minimum lethal dose: defined by Haag and Ambrose, 1937 as the smallest dose that kills 60% of the animals tested

the animals died after just 1–2 days, and only liver enlargement was observed. Down to repeated doses of 5 ml/kg (3–30 times), thrombi were observed in the veins of the renal cortex in 60–70% of the animals. In general there was enlargement of the liver, histological examination revealing vacuolar degeneration of the liver cells especially around the central veins and, in the kidney, vacuolar degeneration and necrosis of the tubular cells. 2 and 1 ml/kg were tolerated up to 28 times (Harris, 1949). Geiling et al. (1937, 1938) administered "pure" (not further specified) DEG by stomach tube to rats (breed not stated) of both sexes at doses of 0.5 ml/kg 3 times daily, (24 days), 2.25 ml/kg 3 times daily, (8 days), 3.0 ml/kg 3 times daily, (7 days) or 4.0 ml/kg 3 times daily, (5 days). The number of animals per group was between 2 and 5. Occurrence of symptoms of intoxication was dose-dependent. After repeated administration of 0.5 ml/kg there was nothing unusual in the animals' behaviour. Repeated administration of 2.25–4.0 ml/kg resulted in rough fur, increased thirst, diuresis and refusal to eat, starting from the 4th dose; later there was increased rapid breathing, followed by coma and death within 2–5 days. 20% DEG (purity not stated) in the feed caused the death of female rats (no information on breed, number per group or other parameters) within 2 weeks. Deaths also occurred in this time at 10 and 5%. Concentrations of 5 and 4% DEG in the drinking water led to death of all animals used within 8 days, whilst 0.25% caused delayed growth. 0.125% had no effect. Precise test details and pathological-histological organ changes are not given (Holck, 1937). To clarify the question of the extent to which mechanical alteration of the bladder wall by calcium oxalate crystals and bladder stones can be causally responsible for development of the bladder tumours described in earlier tests, a 28-day feeding test was conducted, with special reference to the excretion of oxalic acid in the urine. Five male and five female Wistar rats each received 500, 2500 or 10,000 ppm DEG in the feed, 10 male and 10 female rats received 40,000 ppm, and a similar number of animals served as controls. Five males and five females of the 40,000 ppm group and of the control were observed for 3 weeks. The following parameters were investigated: clinical symptoms of intoxication, intake of feed, weight gain, mortality, blood, clinical-chemical parameters, plasma enzyme activities, urine, in particular the concentration and quantity of oxalic acid in the urine, and pathology and histology of the major organs. In the 40,000 ppm group, a significant increase in the concentration of oxalic acid (mg/l) and in the overall amount of oxalic acid (µg/16 hours) was seen in the urine of male and female rats; Calcium oxalate crystals in the urine were only observed in the males. On discontinuation of treatment, calcium oxalate excretion was reversible. In the 500, 2500 and 10,000 ppm dose groups, there was no definite increase in the concentration or amount of oxalic acid.

Apart from a significant reversible decrease (in the 3-week follow-up period) of absolute brain weights of females in the 40,000 ppm group, there was nothing unusual in any of the other parameters investigated. This observed decrease of brainweight was not considered by the authors to be treatment-related (BASF, 1988). Two female and one male rabbit (breed unknown, weight between 2.4 and 2.7 kg) received DEG orally by stomach tube twice daily at doses of 0.5 ml/kg (1 animal 20 times) and 1.0 ml/kg (2 animals each 8 times). Symptoms of intoxication occurred after 36–40 and 72 hours: general weakness, increased breathing, anuria, coma after about 60 hours (8 times 1 ml/kg) or 213 hours (20 times 0.5 ml/kg) and finally death (Geiling et al., 1937). Six rabbits were given 1–4 ml DEG/kg (degree of purity not stated) in the form of 1–5% DEG solutions as drinking water over 5–28 days. One rabbit died after intake of 1 ml DEG/kg for 7 days (kidney damage, pulmonary oedema). $^4/_5$ rabbits, killed after 28 days, exhibited kidney lesions (vacuolar degeneration, necrosis and calcification of the tubules), and 2 of them also exhibited liver lesions (vacuolar degeneration) (Kesten et al., 1939). When DEG was administered orally by stomach tube to 5 guinea pigs at doses of 2–5 ml/kg over 2–12 days, there were kidney lesions in all 5 animals, and liver lesions (vacuolar degeneration) in 2 of them (Kesten et al., 1939). Repeated oral administration (3 times, 1.5 ml/kg daily on 2 consecutive days, total dose 9 ml/kg) led to the death of 4 out of 6 dogs (breed not stated) after 2–4 days. Sometimes the animals vomited severely; the cause of death was stated to be liver and kidney damage, which was not described in more detail (Hanzlik et al., 1939). One dog (breed unknown, weight 11.3 kg, age 10 months, male) was administered 1.5 ml of "pure" DEG/kg three times daily (no further details given) by stomach tube up to a total dose of 9 ml/kg. The first symptoms of intoxication were observed after 20 hours: weakness, increased breathing rate, diuresis and vomiting. Then followed anuria, convulsions, coma and finally death after 86 hours (Geiling et al., 1937). In a preliminary test, 6 cats were administered undiluted DEG (commercial product) by stomach tube in doses of 2–10 ml/kg, 1–13 times; liver and kidney lesions were reported (no further details, Harris, 1949).

7.3 Skin and mucous membrane effects

In a preliminary study, 25 ml/kg DEG (degree of purity not stated) was applied to the shaved abdominal skin of 5 rats anaesthetized with ethyl urethane (breed not stated), once for 2 hours and twice daily on 2 days for 2 hours each time. In 5 non-anaesthetized guinea pigs, 25 ml/kg of the test substance was applied once for 2 hours or 4 hours on the damaged skin (no further details) or for 2 hours, twice daily for 14 days, on the intact skin. No changes appeared on the skin of the animals of either species after single or multiple application of DEG, and there were no signs of inflammatory irritation, erosion or scab-formation (Loeser, 1954). In groups of 5 rats, no local changes were seen in the tail skin after their tails had been immersed in undiluted DEG for 4 hours daily for 4 days or for 1 or 4 hours daily for 6 days (Loeser, 1954). Undiluted DEG, instilled into the conjunctival sac of rabbits, dogs and cats (no details of quantity), did not cause any visible changes; corneal reflexes and pupillary reaction were normal (Loeser, 1954). In the rabbit eye, 0.5 ml of undiluted DEG caused little or no irritation of the mucous membranes (Carpenter and Smyth, 1946). No inflammatory changes or other abnormalities were observed after applying DEG to the oral mucosa of dogs, rabbits and rats (Loeser, 1954). Introduction of the test substance into the stomach of the dog did not cause vomiting (no further details, Loeser, 1954).

The above tests for local effects of DEG on various mucous membranes are of a preliminary nature. No details are given regarding purity of the substance, breeds of animals used, dose applied, duration of the effect or group size. Intracutaneous (0.2 ml), subcutaneous (0.2 ml, 0.8 ml), intramuscular (0.2 ml) and intraperitoneal (rat, mouse: 0.2 ml) injection of undiluted DEG (purity not stated) and intravenous injection (dog, cat: 1.0 ml of 50% DEG, solvent not stated) did not cause lesions at the injection sites in the animals tested (rat, mouse, rabbit, dog, cat), leading the authors to conclude that there is good local tolerance (Loeser, 1954).

7.4 Sensitization

No information available.

7.5 Subchronic and chronic toxicity

For investigation of histopathological changes in the central nervous system after repeated action of DEG, sexually mature mice and rats (no information on breed or sex of the animals or on the allocation and number of animals in the test groups) were given DEG (purity not stated) orally by stomach tube over a period of 3 to 7 months at a daily dose of 300 mg/kg. In a second experiment the animals inhaled 5 mg/m^3 daily for 3 months (daily duration of inhalation not stated), and in a third study 2.8 g/kg was applied daily to mice by the epicutaneous route (precise details not given). At the end of each test the animals were decapitated, dissected and examined histologically. In all treated groups, there was dose-dependent oedema and hyperaemia both in the brain and in the spinal cord, and localized tissue bleeding occurred; to some extent destruction of neurones was observed, with compensatory outgrowth of glial cells. As there is insufficient documentation of the test set-up and results, only limited evaluation of the findings is possible (Marchenko, 1973). Groups of 12 rats (6 male, 6 female, breed not stated) were given 1, 2, 5, 10 or 20% aqueous solutions of "pure" (no additional information) DEG as drinking water. 12 animals receiving distilled water served as controls. The average weight of the rats was 100 g when the test began. Treatment lasted for 3 months. The rats in the groups which received the 5, 10 and 20% solutions of the test substance exhibited, in the course of 3–5 days, rapid weight loss, stupor, increasing weakness and reduced reaction to external stimuli. There was also haemoglobinuria and severe diuresis, followed by exsiccosis. Splenatrophy was evident on dissection. Microscopically, the liver cells were reduced in size and more densely packed and in the kidney the tubular epithelium was severely swollen and vacuolized, with sporadic necrosis (Loeser, 1954; Bornmann, 1954a, b, 1955). In another experiment, 10 ml/kg of a 1, 5, 10 or 20% aqueous solution of DEG was administered by stomach tube 5 times weekly over a period of 50 days (approx 0.1, 0.5, 1.0, 2.0 g/kg) to groups of 6 male and 6 female rats (starting weight approx 50 g). The controls received 10 ml/kg distilled water. Weight gain was normal in all groups during the test. There was no difference in mortality between treated and control rats. Concurrent deaths and diseases were attributed by the authors to technical errors in administration. On dissection, no evident effects were diagnosed in any animal, the weights of the endocrine glands (hypophysis, thyroid, adrenal glands, ovaries) corresponded to the norm and histologically the heart, liver, stomach, intestine, bladder, spleen and endocrine glands were unchanged (Loeser, 1954).

A further experiment investigated the possible toxic effects to rats (initial weight 120–130 g) of administering 5 and 10 ml/kg of a 50% solution of DEG (2.5 and 5.0 g/kg) by stomach tube, twice weekly over a period of 6 months. The two test groups each comprised 45 females and 45 males. The control group comprised 60 animals (30 male, 30 female) which were given 10 ml distilled water/kg. No impairment was observed in the rats treated with DEG in comparison with the control group, in terms of weight gain, general condition and mortality. On dissection (after 6, 12 and 18 weeks and at the end of the test), no pathological changes were observed, and the fine structure of the organs examined (kidney, bladder, liver, spleen, endocrine organs) was described as normal (Loeser, 1954). When groups of 5 rats were given a DEG dose of 2.5 ml/kg by stomach tube (undiluted), three times daily, or 7.5 ml/kg once daily, they died in 11–50 days; apart from a few exceptions, the liver and kidneys exhibited vacuolar degeneration (Weatherby and Williams, 1939). Haag and Ambrose (1937) gave white rats (breed and sex not stated) an aqueous solution of DEG (degree of purity unknown) in concentrations of 0.03, 0.1, 0.3, 1, 3 and 10% as drinking water for 100 days. The control group received only distilled water. Each test group comprised 5 animals. At the end of the test, all the surviving rats were killed and examined macroscopically. Weight gain was normal in all groups. At 10 and 3% DEG in the drinking water, all animals died after 8 or 9 days. In the other groups, a maximum of 1 to 2 animals died during the test. On dissection, all organs were normal, apart from the spleens of the animals which received 1% solutions of DEG. The changes in the spleen (deposition of pigment, staining for iron negative) were not described in more detail. The deaths that occurrred concurrently in the treated groups were ascribed by the authors partly to the greatly reduced water intake of the animals. No further information was given. Morris et al. (1942) administered DEG (degree of purity unknown) in the feed of inbred albino rats at doses of 1.71% and 3.42%. Each test group and the control group comprised 10 animals (6 male, 4 female). The rats were kept in individual cages, with body weight and food intake determined at weekly intervals. The study ended after 24 months, and all animals still alive at this point were put to death. During the first 55 weeks of testing, food intake and weight gain were normal in all groups. After treatment with DEG, bladder stones consisting mostly of calcium oxalate were found in 3 males (test group not stated), and were accompanied by chronic cystitis. In addition, oxalate crystals were seen in the tubules. Furthermore, there was kidney damage (atrophy of the tubules, infiltration of lymphocytes, fibrosis) and changes in the liver (diffuse and centrilobular atrophy, fatty degeneration, bile-duct proliferation). Assignment of the observed changes to the different animals and test groups is not given, therefore the possible dose dependence of the findings cannot be assessed. Groups of 15 male and 15 female rats received DEG in the feed at concentrations of 0, 0.4, 2.0 and 4.0% for 90 days. In a second experiment groups of 10 male and 10 female rats received DEG in the feed at concentrations of 0, 0.85, 0.17, 0.4 and 2.0% for 225 days. The same DEG sample, containing less than 0.01% ethylene glycol, was used in both tests. 4% (90 days) in the feed caused the death of 6 male rats with kidney damage. The surviving animals in this group exhibited reduced weight gain, increased water intake, increased production of urine, signs of blood thickening, enlarged kidneys and, histologically, kidney and liver lesions. Concentrations of 0.4 and 2.0% DEG (225 days) in the feed led to excretion of oxalate crystals in the urine and slight impairment of kidney function (reduced concentrating ability). At 0.17% DEG in the feed there was 13–23% greater excretion of oxalate crystals in the urine of male rats. Effects were no longer observed with a diet containing 0.08% DEG. On the basis of these results, an acceptable daily intake of 38 mg/day was established for a man weighing 70 kg, using a safety factor of 100 (Gaunt et al., 1976). In a drinking test, 5 male rats each received 1% DEG (no information on purity) in the drinking water (approx 2.29 mg/kg/day) or 0.79% (approx 1.17 mg/kg/day) over a maximum period of 24 months. Apart from considerable fluctuation in food intake during the test period, histopathological examination revealed calcium oxalate calculi in the bladder, and kidney lesions (Hanzlik et al., 1947).

Groups of 10 male and 10 female rats received DEG (purity not stated) in the drinking water at concentrations of 1% and 0.3% (males 2.35 and 0.66 g DEG/kg/day; females 1.94 and 0.59 g DEG/kg/day) over 175 days. Intermediate dissection being conducted on 5 animals per group after 110 days. Corresponding controls were also used. Only 2 rats from each group died during the first 100 days. There were no deaths when treatment was continued for 75 days. No changes were seen in the internal organs on histopathological investigation (Weatherby and Williams, 1939).

In another test, rats (no data on breed or sex) received DEG (purity not stated) in the drinking water; 17 rats 0.5% (approx 0.24 ml DEG/kg/day) DEG in the drinking water over 33–124 days; 30 rats 1% (approx 0.6 ml DEG/kg/day) over 33–174 days; 25 rats 3% (approx 3.5 ml DEG/kg/day) over 15–95 days; 35 rats 5% (approx 6 ml

DEG/kg/day) over 1–6 days). All animals survived DEG concentrations of 0.5 and 1% in the drinking water without symptoms of intoxication or damage to the organs; 3% caused the death of 14/25 rats after 5–56 days; 5% led to the death of 9/35 rats after 1–6 days. In the animals in the latter two test groups (3% and 5% DEG in the drinking water), which died or were killed at various times, extensive epithelial lesions were found in the renal tubules with retention of urine, increased residual nitrogen and uraemia. Vacuolar degeneration was found in the liver and adrenocortical cells (Kesten et al., 1937).

Three dogs received 7.5 ml undiluted DEG/kg daily, by stomach tube. They died after 4–13 days, with kidney and liver lesions (vacuolar degeneration) (Weatherby and Williams, 1939).

7.6 Genotoxicity

In a modified Ames test with Salmonella typhimurium strains TA 98, TA 100, TA 1535 and TA 1537, at concentrations of 62.5, 250 and 1000 µg/plate without addition of a metabolizing system, DEG was negative for all the strains used (no precise details, Pfeiffer and Dunkelberg, 1980). DEG was investigated in the Ames test (TA 98, TA 100, TA 1535, TA 1537) with and without S9-mix, at concentrations of 1, 3, 10, 30 and 111.8 mg/plate. There was no increase in the number of revertants (Slesinski et al., 1986; Hengler and Slesinski, 1984). In CHO cells, at concentrations of 30 to 50 mg/ml with and without S9-mix, no chromosome aberrations were found (Slesinski et al., 1986; Guzzi and Slesinski, 1984). Moreover, there were no indications of genotoxic activity in a CHO/HGPRT test and in an SCE test, at concentrations of 30 to 50 mg/ml with and without S9-mix (Slesinski et al., 1984, 1986).

7.7 Carcinogenicity

In a chronic-static inhalation test on white mice (weight 18–19 g at start of test), an aerosol-vapour mixture produced at 30° C to 35° C was supplied in a 100-liter cage (DEG evaporated from a Petri dish heated to 30–35° C in the middle of the cage). 16 female mice inhaled concentrations of DEG (degree of purity unknown) in the range 0.004–0.005 mg/l (4–5 mg/m^3, about 0.92 ppm) for 2 hours daily over a period of 7 months. The manner of calculation and determination of concentration was not stated. The control group comprised 20 animals of the same weight and sex. The treated animals exhibited bronchitis and interstitial pneumonia. In the kidneys, dystrophy of the epithelium and round-cell infiltrations were described, and slight protein dystrophy was diagnosed in the liver.

Out of 12 treated animals, 10 developed a tumour 2.5 to 11 months after the end of the experiment. There was a lymphosarcoma in the back of the neck of one mouse, in another animal a smooth-cell, non-keratinizing tumour of the mammary gland was found, and adenocarcinomas of the mammary glands were observed in 7 mice; another animal had a solid tumour, which was not further defined. None of the control mice exhibited neoplastic changes on dissection (Sanina, 1968). The study does not meet current requirements (no information on the purity of the DEG or on the manner of determination of concentration and analysis of the aerosol-vapour mixture; inadequate documentation of the results). 74 mice (F1 C57xCAB, sex not stated) each received 2 drops of undiluted DEG on the shaved skin between the shoulder blades, 3 times weekly over a period of 2 years. After 2 years, a histologically confirmed papilloma was only found in 1 treated mouse. No difference from the controls was found (Vasil'eva et al., 1971). On adding DEG at 2% to cigarette tobacco as a humectant, the incidence of skin tumours in mice which had been traeted on the dorsal skin with smoke condensate containing DEG (0.7 ml/dose, 3 times weekly), was not significantly different from the controls (smoke condensate without addition of DEG) (Dontenwill et al., 1970). In another carcinogenicity test, 42 female and 57 male rats were given 0.84 g/kg by stomach tube, 3 times weekly for a total of 243 doses. In addition, 42 female and 41 male rats were treated subcutaneously with 2.5 g/kg once a week for a total of 82 doses. 32 female and 32 male rats were used as untreated controls. Furthermore, 54 female and 36 male mice were treated 198 times with an oral dose of 0.42 g/kg 3 times weekly, and 54 female and 39 male mice were treated 66 times with a subcutaneous dose of 1.25 g/kg once weekly. 52 female and 30 male untreated mice served as a control. According to the authors the oral dose corresponded in each case to one hundredth of the LD_{50} (for summary see Table 2). No reasons were given for the doses selected (Maksimov et al., 1983). In the rats treated orally, the percentages of all tumours (34.8%) and of all haemoblastoses (13.6%) were higher than for the untreated controls (18.3% and 3.3% respectively). With subcutaneous administration the values were 64.7% and 47%, against 18.3% and 3.3% for the controls ($p < 0.05$). Moreover, occasional lung tumours were found in the rats treated by either route, but not in the controls. In mice too, after oral and subcutaneous administration, the percentages of all tumours and of haemoblastoses were higher than in controls. In conjunction with the observation that lung tumours only occurred in the treated rats and that the tumours observed in mice developed earlier than in the control group, the authors concluded that DEG has a slight carcinogenic action (no information on purity of the DEG, or on the manner of evaluation and frequency of tumours in the historical controls; furthermore, in each case, only 1 dose was used, so that dose/effect relationships cannot be determined; there is no information on preliminary tests for determining the maximum tolerated dose (MTD, Maksimov et al., 1983). Therefore the study does not meet current requirements.

Groups of 12 male Osborne-Mendel rats were given 1, 2 and 4% DEG (purity not stated) in the feed. The control group also contained 12 animals. The rats were kept in individual cages, and weights and food intake were determined at weekly intervals. Test duration was 2 years, then all the surviving animals were killed. No female rats were used (Fitzhugh and Nelson, 1946). During the first 52 weeks of the experiment, the rats in the highest dose group showed a significantly ($p = 0.05$, test method not stated) reduced weight gain compared with the control animals. There were no differences in food intake between treated groups and controls. Five animals in the control group survived to the end of the test, but none in the highest dose group. For the two dose groups with 2 and 1% DEG in the feed, no increased mortality was found in comparison with the controls. DEG produced dose-dependent chronic-toxic lesions in the lower urinary tract, which are described below. In addition, however, there were bladder stones (low dose 2, medium-dose 7, high dose 11) and tumours of the bladder (medium dose 6, high dose 5, for summary see Table 3). The first bladder stone appeared in the high dose group after 32 weeks, and the first tumour of the bladder was observed in the same group after 53 weeks. In 10 animals with a bladder tumour, a stone was also found. The bladder stones consisted of calcium oxalate. The bladder tumours grew primarily papillomatously into the lumen, though there was also intramural growth. The neoplastic changes exhibited different stages of malignancy. A metastasizing carcinoma developed in one case. In animals which received 4% DEG in the feed, on dissection the kidneys were enlarged and hydronephrotic, and microscopic examination revealed focal tubular and glomerular atrophy, formation of hyaline casts, hydropic degeneration and calcification. In the animals of the 2% and 1% dose groups these changes were less pronounced, their extent and frequency being dose-dependent. In the livers of the rats (only medium and high

Table 2. Carcinogenicity of DEG (Investigations by Maksimov et al., 1983)

Species	Route of administration	Number of animals	Dose	Number of tumours	Haemoblastoses	Mammary tumours	Lung	NN	S.c. tumours	1st tumour after month	Mean life-span
Rat	oral	f 42 m 57 99	243 × 0.84 g/kg 3 × per week	19 4 23 = 34.8%	5 4 9 = 13.6%[a]	13 0 13 = 35%	1 0 1	2 0 2		10	17.8
Rat	s.c.	f 42 m 41 83	82 × 2.5 g/kg 1 × per week	19 25 44 = 64.7%[a]	10 22 32 = 47%[a]	11 0 11 = 34%	2 0 2		3 0 3	15.5	22.6
Untreated controls		f 32 m 32 64	—	10 1 11 = 18.3%	1 1 2 = 3.3%	9 0 9 = 30%				11	20
Mouse	oral	f 54 m 36 90	198 × 0.42 g/kg 3 × per week	20 15 35 = 39.3%[a]	8 7 15 = 16.8%[a]	10 0 10 = 18.9%	3 8 11 = 12.4%			2	15.2
Mouse	s.c.	f 54 m 39 93	66 × 1.25 g/kg 1 × per week	21 13 34 = 37.4%	4 3 7 = 7.7%[a]	15 0 15 = 27.8%	3 6 9 = 9.9%			2.5	14.5
Untreated Controls		f 53 m 33 86	—	11 2 13 = 16.2%	2 1 3 = 3.7%	6 0 6 = 11.5%	3 1 4 = 4.9%			8	15

[a] $p < 0.05$

The percentages were taken from the original work; sometimes they cannot be reconstructed.

Diethylene glycol 235

Table 3. Testing of DEG for carcinogenicity over a period of 2 years, on male Osborne-Mendel rats (Fitzhugh and Nelson 1946)

Number of male animals	Dose % in feed	Delayed weight increase	Increased mortality compared to controls	Bladder calculi	Bladder tumours	Kidney damage	Liver damage
12	1	Slight	None	2	0	Slight	Rare
12	2	Slight	None	7	6	Moderate	Slight
12	4	Marked	Significant	11	5	Marked	Moderate
12	–	–	–	–	–	–	–

dose group), hydropic degeneration and focal cell necrosis were diagnosed. The control animals examined had neither tumours nor calculi in the bladder, and the kidneys and liver were clear of pathological changes (Fitzhugh and Nelson, 1946).

Weil et al. (1965, 1967) conducted a carcinogenicity study on male and female rats (Carworth Farms Nelson). Each test group comprised 15 to 20 animals. DEG (commercial product, contaminated with 0.031% ethylene glycol) was administered in the feed for a period of 2 years at doses of 0% (control), 2% and 4%, to rats of different ages (rats weaned from the mother, 2 months old and 1-year-old animals). Supply of DEG in the feed was adjusted to the age of the animals, so that all the animals received roughly equal amounts of DEG. None of the male rats included in the test at 1 year survived, whereas all the female rats were still alive. The rats included in the test at 2 months and after weaning also survived. At 4% DEG in the feed (age-adjusted), bladder stones developed only in male rats of all age groups (14–25% of the animals), and a bladder tumour was only found in 1 male rat (group of weaned rats).

Table 4. Summary of 2-year feeding test with 4% DEG in the feed (Weil et al., 1965) Occurence of bladder calculi or bladder tumours

Number of rats	Dose (%)	Bladder calculi	Bladder tumours
20 male	4	8	1
20 female (weanlings)	4	0	0
20 male	4	5	0
20 female (2 months old)	4	0	0
20 male	4	5	0
20 female (1 year old)	4	0	0

At 2% DEG in the feed, neither bladder stones nor bladder tumours were found, just as in the controls. Feed with 4% DEG (age-adjusted) was supplied for a period of 90 days to 19 male rats newly weaned from the mother, and neither bladder stones nor bladder tumours developed, probably because the duration of administration was too short (Weil et al., 1967). Other tests (Weil et al., 1965, 1967) investigated whether bladder tumours can also develop in the absence of bladder stones or glass beads without feeding DEG, or only if a sham operation is performed on the urinary bladder. For this purpose the following tests were conducted:

- As a result of feeding DEG in several test arrangements over a period of 2 years (4% in the feed, to 136 male rats separated from the mother), bladder stones detectable by fluoroscopy were found in 74% (58/78) of the rats given 6 g/kg and 31% (18/58) of the rats given 4 g/kg, and bladder tumours in 2.6–3.4% of the animals; the stones were obtained on dissection.
- In another long-term test, 118 male rats received 4% DEG in the fed in the first 6 months, and then 6%. In 46 rats in which stones were found and surgically removed, dissection after death revealed bladder stones in 94% and tumours in 4%; on the other hand, in 72 rats which had not been operated on, there were only 24% bladder stones and 3% tumours. In the first operated group, DEG was no longer supplied after surgery, whereas the unoperated group continued to receive DEG until the end of their life (Weil et al., 1967).
- After a sham operation on the bladder of 21-day-old rats (20 male, 10 female treated animals, 18 male and 9 female controls), DEG was administered (4% in the feed, age-adjusted) over a period of 2 years. Bladder stones developed in 75% of the males (15/20), and bladder tumours in 10% (2/20), whereas neither bladder stones nor tumours were seen in females fed with DEG after the same preliminary treatment. When a sham operation was conducted without subsequent feeding of DEG, again neither bladder stones nor tumours developed in male or female control rats (for summary see Table 5) (Weil et al., 1967).
- In other tests, without feeding DEG, the bladder stones (calcium oxalate calculi obtained from the various investigations mentioned above, or glass beads, were surgically inserted into the bladder of untreated male and female rats (Weil et al., 1965, 1967). For comparison, an additional group of rats underwent a sham operation without implantation of a foreign body in the bladder. The results of these investigations are presented in Table 6.

In the 7 rats (4 male, 3 female) which developed bladder tumours (papillomas) after implantation of a bladder stone from DEG donor animals, there were always several bladder stones. It followed from these investigations that a sham bladder operation alone or a bladder operation with implantation of glass beads or calculi was

Table 5. Summary of the sham operation on the bladder of 21-day old animals and subsequent feeding of DEG (4%, age-adjusted) (Weil et al., 1967)

DEG concentration in the feed	Number and sex of rats	Rats with calculi	Rats with tumours
4%	20 male	15	2
4%	10 female	0	0
Control (sham operation without feeding of DEG)	10 male 9 female	0	0

Table 6

Pretreatment no DEG feeding	Number of usable rats	Rats with bladder calculi	Rats with bladder tumours
Implantation of 1 bladder stone in each case from DEG donor animal	37 male 30 female	18 (49%) 8 (27%)	4 (11%) 3 (10%)
Implantation of 1 glass bead in each case	50 male 27 female	14 (28%) 5 (18.5%)	1 (2%) 0 (0%)
Sham operation	16 male 8 female	5 (31%) 4 (50%)	0 (0%) 1 (12.5%)

sufficient to produce bladder tumours and that additional feeding of DEG is not necessary for the development of bladder tumours. The incidences of bladder tumours and of all tumours in the three different test groups were not statistically different. In the tests mentioned above (Table 5), 19 male and female rats just separated from the mother did not develop bladder stones or tumours after the sham operation without DEG. In contrast to this, in the test summarised in Table 6, bladder stones developed in male and female rats after a sham operation without feeding of DEG and one of the 8 females even had a bladder tumour. In no case was a bladder tumour found if the foreign body had not been detected in the bladder previously or at the same time, or a sham operation had not been undertaken. Analysis of the bladder stones showed that after feeding of DEG, bladder stones found in the rat bladder contained on average 86.2% calcium oxalate, whereas the bladder stones that formed after implantation of glass beads or after performing a sham operation had calcium oxalate contents of less than 2%. Analysis by X-ray diffraction showed that these stones were composed of calcium, magnesium and ammonium phosphates (Weil et al., 1965). According to these investigations, development of bladder stones after feeding of DEG was dose-dependent but not concentration-dependent. Bladder tumours did not develop unless foreign bodies were present in the bladder beforehand. In the opinion of the authors, the observed bladder tumours were probably the result of mechanical irritation or operative trauma. Bladder stones developed more quickly in 1-year-old rats than in young rats separated from the mother (Weil et al., 1965, 1967).

7.8 Reproductive toxicity

Wegener (1953) investigated the effect of repeated administration of pure DEG (no further details) on the reproductive capacity of the rat. 10 male and 10 female albino rats were administered 1.0 ml/100 g body weight of a 20% aqueous solution of DEG (approx 2 ml/kg daily for 8 weeks, by stomach tube. The control group (10 males, 10 females) received 1.0 ml distilled water/100 g body weight. The animals of the P-generation that had been pretreated in this way were paired within each group when the weight of the females was 170–215 g, changing the males daily. Five females received DEG until they gave birth, and the other five until the young had been weaned. Total duration of the test was >12 weeks. Growth of the F1 generation was monitored from the 15th to the 60th day of life by checking their weight at intervals of 2 days. When the young reached a body weight of 60 g they were separated from the mother. Onset of oestrus was determined for the females by daily vaginal smears. At the age of about 100 days, a proportion of the F1 generation was killed. Weights of the pituitary, thyroid, adrenal glands,ovaries and testes were determined, and these organs were also examined histologically. 10 males and 10 females from each F1 group were mated when the average weight of the females was 200 g. Weight curves were also constructed for the F2 young, onset of sexual maturity was determined, and after the animals were killed the aforementioned organs were weighed and examined histologically. Treatment of the P-generation with DEG for 12 weeks did not impair reproduction. The test animals and the controls became pregnant at almost the same time, litter size averaged 8–10 young, and the young exhibited similar, uniform development. Growth and onset of oestrus were not disturbed. The endocrine glands investigated showed no differences from the controls with regard to weight and fine structure. The receptiveness and litter size of the untreated F1 generation were the same as those of the P-generation, and the F2 generation was normal with regard to weight gain, onset of sexual maturity, and weight as well as histology of the organs examined. At the doses used in this study there was nothing to suggest impairment of reproduction or any embryotoxic action of DEG (Wegener, 1953). Three groups (each comprising 30 male and 30 female Sprague-Dawley rats) received DEG in the drinking water at doses of 150, 500 and 1500 mg/kg/day before mating and thereafter to the end of the test. An appropriate control was conducted. 15 pregnant females from each group were killed on the 20th day of gestation, and the foetuses were examined (for skeletal or visceral malformations);the other 15 pregnant females gave birth at the normal time. DEG did not affect the survival rate, clinical behaviour or water intake of the parental generation. At 500 and 1500 mg/kg/day, weight gain in the male rats was reduced during the first week. The relative kidney weights for the males of the parent and F1 generation were increased, at 1500 mg/kg/day. DEG affected neither reproduction in the parents nor development (no malformations), survival or growth of the F1 generation (Rodwell et al., 1987). After injecting 0.05 and 0.1 ml of undiluted DEG in the yolk sac of hens' eggs before the beginning of incubation, there was a reduction in the hatching rate by 20 and 45% respectively. No abnormalities were observed in the hatched chicks (McLaughlin et al., 1963).

7.9 Effects on the immune system

No information available.

7.10 Other effects

No information available.

8. Experience in humans

In 1937 there was mass poisoning by the medication "Elixir Sulfanilamide" (manufactured by Massengill Co.), which contained 72% DEG as a solvent for sulphonamide (Geiling et al., 1937; Geiling and Cannon, 1938; Lynch, 1938; Calvery and Klumpp, 1939). Symptoms of poisoning after ingesting this preparation were described for 105 deaths and in a further 248 people, though it could not be excluded in every single case that the underlying disease being treated with the medication was jointly responsible for the symptoms. Repeated intake of the DEG-containing preparation led to dizziness and vomiting, and often headaches. Unless the medication was withdrawn, these symptoms were followed by polyuria, oliguria and finally anuria. Abdominal and back pains were often described. Prior to death there were disturbances of consciousness, coma and, in some children, tremors and convulsions, which are interpreted as a consequence of uraemia. If laboratory data were recorded (albuminuria, casts and erythrocytes in the urine, increased residual nitrogen in the blood), they confirmed progressive kidney failure. For 96 patients, death was attributed to ingestion of the medication. On average they died after 9.4 ± 3.4 days. For the other 9 deaths the causal relationship was not certain. 200 adults and 48 children survived repeated ingestion of the preparation. The doses used were up to 340 ml (corresponding to 245 ml DEG). On average, doses of 44.2 ml (corresponding to 32 ml DEG) were survived by children up to 14 years and of 83.7 ml (corresponding 60 ml DEG) by persons of 15 years and older. The amounts of the substance actually absorbed cannot be assessed, because ingestion of the medication was often soon followed by gastro-intestinal disorders, which led to discontinuation of the preparation (Calvery and Klumpp, 1939). A histological report is available for 12 intoxications resulting in death. Autopsy revealed generalized oedema (ascites, hydrothorax, hydropericardium, pulmonary oedema). In addition in some cases there was bleeding into the gastro-intestinal tract and lungs. The increased capillary permeability can be correlated with the clinical picture of acidosis and uraemia. The principal signs of intoxication were lesions in the kidneys and liver. There were cortical necroses and infarcts in the enlarged kidneys. The main microscopic finding was nephrosis with severe vacuolization of the tubular epithelium (hydropic degeneration). Both hyaline casts and accumulations of erythrocytes and leucocytes were described in the collecting tubules. As a rule the livers were enlarged, pale and of soft consistency. The picture corresponded to chronic congestion of the liver. Centrilobular hydropic degeneration was observed microscopically. Fatty degeneration was only observed at the margins of the liver. Signs of inflammation (infiltration of leucocytes) were only reported in 2 cases. The changes found in other organs could all be explained by the uraemia and terminal cardiac failure (Geiling and Cannon, 1938). In a report from South Africa concerning 7 children, a common cause of intoxication was shown retrospectively to be the use of sedatives in which DEG was used as solvent instead of propylene glycol. The picture of intoxication corresponded to that described for adults (see above). The following findings were recorded (frequency): feverish prodromal stage (7), vomiting (7), anuria (7), diarrhoea (3), dehydration (7), metabolic acidoses (7), hepatomegaly (7), disturbances of consciousness (5), enlargement of the kidneys (3), leucocytosis (7) (14,000–29,000/mm^3), increased blood urea (7), increased SGPT with normal bilirubin (7), neurological symptoms (4). In all cases death occurred unexpectedly and suddenly through cardiac failure. The findings at autopsy were also similar to those reported for adults (see above), in particular, pathological lesions of the kidneys and liver. No pathological changes in the brain were diagnosed, even in the 4 cases which previously had neurological symptoms. No statements were made concerning the absorbed doses of DEG (Bowie and McKenzie, 1972). Another case has been reported concerning intoxication of a 65-year-old alcoholic after ingestion of about 150 ml of pure DEG. The man became comatose 16 hours after ingestion, necessitating artifical respiration. The patient also developed severe metabolic acidosis with high oxalate level in the urine. After plasma alkalization and peritoneal dialysis the patient survived the poisoning without consequences, according to a follow-up investigation a year later (Auzépy et al., 1973). Telegina et al. (1971) reported a retrospective epidemiological study, without controls, on workers who had contact with DEG in the extraction of benzene and its homologues; no details of exposure levels were reported. After an exposure period of 1–9 years, no indication of an increased cancer risk was found in 90 workers (56 men, 34 women, age 20–49 years). Because of the inadequate methodology of the investigation, few conclusions can be drawn from this study. 15 ml DEG was tolerated by humans (Remmer, 1985). A figure of 30–60 mg is given as the safe daily dose for humans (Remmer, 1985; Altmann et al., 1986). No information is found in the literature concerning the occurrence of bladder stones in humans after ingestion of DEG.

References

Altmann, H.J., Grunow, W., Krönert, W., Uehleke, H.
Gesundheitliche Beurteilung von Diethylenglykol in Wein
Bundesgesundheitsblatt, 29, 141 (1986)

Auzépy, P., Taktak, H., Toubas, P.-L., Deparis, M.
Intoxications aigues par l'éthylène glycol et la diéthylène glycol chez l'adulte
Semaine des Hôpitaux de Paris, 49, (19), 1371–1374 (1973)

BASF AG
Toxicology department.
Tests on the oral toxicity of diethyleneglycol to rats.
4 wk dietary administration and 3 wk observation period.
Project No. 3050036/8526, report dated 18.4.1988 and Supplement dated 15.9.1988.
Commissioned by the Employment Accident Insurance Fund of the Chemical Industry

Bornmann, G.
Grundwirkungen der Glykole und ihre Bedeutung für die Toxizität, 1. Mitteilung
Arzneimittelforschung, 4, 643–646 (1954a)

Bornmann, G.
Grundwirkungen der Glykole und ihre Bedeutung für die Toxizität, 2. Mitteilung
Arzneimittelforschung, 4, 710–715 (1954b)

Bornmann, G.
Grundwirkungen der Glykole und ihre Bedeutung für die Toxizität, 3. Mitteilung
Arzneimittelforschung, 5, 38–42 (1955)

Bowie, M.D., McKenzie, D.
Diethylene glycol poisoning in children
S. Afr. Med. J., 46, 931–934 (1972)

Browning, E.
Diglycol toxicity and metabolism of industrial solvents
Elsevier Publishing Company, Amsterdam, 624–628 (1965)

Calvery, H.O., Klumpp, T.G.
The toxicity for human beings of diethylene glycol with sulfanilamide
South. Med. J., 32, 1105–1109 (1939)

Carpenter, C.P., Smyth, H.F.
Chemical burns of the rabbit cornea
Am. J. Ophthal., 29, 1363 (1946)

Casarett, L.J., Doull, J.
Toxicology.
The basic science of poisons New York, Macmillan Publishing Company Inc., p. 516 (1975)

CRC
Handbook of Chemistry and Physics, C-268 62nd ed., CRC-Press, Boca Raton, Florida, 1981/82

Dontenwill, W., Elmenhorst, H., Harke, H.P., Reckzeh, G., Weber, K.H., Misfeld, J., Timm, J.
Experimentelle Untersuchungen über die tumorerzeugende Wirkung von Zigarettenrauch-Kondensaten an der Mäusehaut
Z. Krebsf., 73, 285–304 (1970)

Durand, A., Auzépy, P., Hébert, J.-L., Trieu, T.C.
A study of mortality and urinary excretion of oxalate in male rats following acute experimental intoxication with diethyleneglycol
Europ. J. Intens. Care Med., 2, 143–146 (1976)

Fitzhugh, O.G., Nelson, A.A.
Comparison of the chronic toxicity of triethylene glycol with that of diethylene glycol
J. Industr. Hyg. Toxicol., 28, 40–43 (1946)

Freundt, K.J., Helm, H., Weis, N.
Activities of urinary enzymes after treatment with 2-ethoxy-ethanol, 2-methoxy-ethanol or diethylene glycol
Tagung der Dtsch. Ges. für Pharmakologie und Toxikologie, R11 (1987)

Gaunt, I.F., Lloyd, A.G., Carpanini, F.M.B., Grasso, P., Gangolli, S.D., Butterworth, K.R.
Studies of the toxicity of diethylene glycol in rats
BIBRA Inform. Bull., 15, 217 (1976)

Geiling, E.M.K., Coon, J.M., Schoeffel, E. W.
Elixir of Sulfanilamide-Massengill Chemical, pharmacologic, pathologic and necropsy reports; preliminary toxicity report of diethylene glycol and sulfanilamide
J. Am. Med. Assoc., 109, 1531–1539 (1937)

Geiling, E.M.K., Cannon, P.R.
Pathologic effects of elixir of sulfanilamide (diethylene glycol) poisoning
J. Am. Med. Assoc., 111, 919 (1938)

Gessner, P.K., Parke, D.V., Williams, R.T.
Studies in Detoxication 86.
The metabolism of ^{14}C-labelled ethylene glycol
Biochem. J., 79, 482 (1961)

Grasso, P.
Review of tests for carcinogenicity and their significance to man
Clin. Toxicol., 9 (5), 745–760 (1976)

Guzzi, P.I., Slesinski, R.S.
Diethylene Glycol:
In vitro cytogenetic testing with chinese hamster ovary cells
Bushy Run Research Center Report 47–95, 11.7.1984
Commissioned by the Union Carbide Corporation

Haag, H.B., Ambrose, A.M.
Studies on the physiological effect of diethylene glycol II. Toxicity and fate
J. Pharmacol. Exp. Ther., 59, 93–100 (1937)

Hanzlik, P.J., Newman, H.W., van Winkle, W., Jr., Lehman, A.J., Kennedy, N.K.
Toxicity, fate and excretion of propylene glycol and some other glycols
J. Pharmacol. Exp. Ther., 67, 101–113 (1939)

Hanzlik, P.J., Lawrence, W.S., Laqueur, G.L.
Comparative chronic toxicity of diethylene glycol monoethylether (Carbitol) and some related glycols: results of continued drinking and feeding
J. Ind. Hyg. Toxicol., 29, 233–241 (1947)

Harris, P.N.
Observations on diethylene glycol poisoning in rats and cats
Arch. int. pharmacodyn., 79, 164–172 (1949)

Hébert, J.-L., Fabre, M., Auzépy, P., Paillas, J.
Acute experimental poisoning by diethylene glycol: acid base balance and histological data in male rats
Toxicol. Europ. Res., 1, 289–294 (1978)

Hengler, W.C., Slesinski, R.S.
Diethylene Glycol: Salmonella/Microsome (Ames) Bacterial Mutagenicity Assay
Bushy Run Research Center Report 47–20, 5.3.1984
Commissioned by the Union Carbide Corporation

Holck, H.G.O.
Glycerin, ethylene glycol, propylene glycol and diethylene glycol
Report on feeding experiments with rats
J. Am. Med. Assoc., 109, 1517–1520 (1937)

Hueper, W.C., Payne, W.W.
Polyoxyethylen(8)stearate, carcinogenic studies
Arch. Environ. Health, 6, 484 (1963)

Karel, L., Landing, B.H., Harvey, T.S.
The intraperitoneal toxicity of some glycols, glycolethers, glycolesters and phthalates in mice
J. Pharmacol. Exp. Ther., 90, 338 (1947)

Kesten, H.D., Mulinos, M.G., Pomerantz, L.
Renal lesions due to diethylene glycol
J. Am. Med. Assoc., 109, 1509–1511 (1937)

Kesten, H.D., Mulinos, M.G., Pomerantz, L.
Pathological effects of certain glycols and related compounds
Arch. Pathol., 27, 447–465 (1939)

Laug, E.P., Calvery, H.O., Morris, H., Woodard, G.
The toxicology of some glycols and derivates
J. Industr. Hyg. Toxicol. 21, 173–201 (1939)

Loeser, A.
Diäthylenglykol
Neuere Beiträge zur Pharmakologie und Toxikologie der Polyglykole
Arch. expl. Path. Pharmakol., 221, 14–33 (1954)

Lynch, K.M.
Diethylene glycol poisoning in the human
South. Med. J., 31, 134–137 (1938)

Maksimov, G.G., Teregulova, O.V., Pylev, L.N., Gilev, V.G., Gubajdullin, R.M.
Untersuchung der blastomogenen Eigenschaften von Diäthylenglykol
Deponirovannye naucnye raboty, VINITI, 1983, Nr. 6801–83, 1–11

Marchenko, S.A.
Pathomorphological changes in the central nervous system under the effect of diethylene glycol
Vrach. Delo, Iss. 2, 138–140 (1973)

Mastromatteo, E.
On the concept of threshold
Am. Ind. Hyg. Assoc. J., 42, 763–770 (1981)

McLaughlin, J., Marliac, J.P., Verrett, M.J., Mutchler, M.K., Fitzhugh, O.G.
The injection of chemicals into the yolk sac of fertile eggs prior to incubation as a toxicity test
Toxicol. Appl. Pharmacol., 5, 760–771 (1963)

Morris, H.J., Nelson, A.A., Calvery, H.O.
Observation on the chronic toxicities of propylene glycol, ethyleneglycol, diethyleneglycol, ethylene glycol mono-ethylether and diethylene glycol mono-ethylether
J. Pharmacol. Exp. Ther., 74, 266 (1942)

Pfeiffer, E.H., Dunkelberg, H.
Mutagenicity of ethylene oxide and propylene oxide and of the glycols and halohydrines formed from them during the fumigation of foodstuffs
Fd. Cosmet. Toxicol., 18, 115–118 (1980)
Personal communication from Professor H. Dunkelberg (11.1.1988)

Remmer, H.
Diäthylenglykol im Wein:
Weniger ein toxikologisches als ein kriminelles Problem
Dtsch. Ärzteblatt, 82, 3165 (1985)

Rodwell, D.E., Davis, R.A., Tasker, E.J., Friedman, M.A.
Fertility and general reproductive performance study in rats with a teratology phase on diethylene glycol
The Toxicologist, 7, 145 (1987)

Rowe, V.K., Wolf, M.A.
Glycols
Patty's Industrial Hygiene and Toxicology, Vol. 20, 3817–3907
eds. Clayton, G.D., Clayton, F.E.
John Wiley & Sons, New York (1982)

Sanina, Y.P.
Long-term consequences of chronic inhalation of diethylene glycol
Gig. Sanit. 33, Iss. 2, 36–39 (1968)

Slesinski, R.S., Guzzi, P.I., Hengler, W.C.
Diethylene Glycol:
In vitro genotoxicity studies CHO/HGPRT mutation test sister chromatid exchange assay
Bushy Run Research Center Report 47–94, 9.7.1984
Commissioned by the Union Carbide Corporation

Slesinski, R.S., Guzzi, P.I., Hengler, W.C., Ballantyne, B.
Evaluation of the cytotoxicity and potential genotoxicity of ethylene glycol and diethylene glycol using a battery of in vitro tests
The Toxicologist, 6, 228 (1986)

Smyth, H.F., Jr., Seaton, J., Fischer, L.
The single dose toxicity of some glycols and derivatives
J. Industr. Hyg. Toxicol., 23, 259–268 (1941)

Telegina, K.A., Mustaeva, N.A., Sakaeva, S.H., Briko, V.I.
Health of persons working with diethylene glycol in the production of aromatic hydrocarbons from crude oil
Gig. Tr. Prof. Zabol., 15 (9), 40–41 (1971)

Ullmann
Encyclopaedia of technical chemistry 4th Edition, Vol. 8, p. 200
Verlag Chemie, Weinheim, 1974

Vasil'eva, N.N., Linnik, A.B., Sgibnev, A.K.
A study on possible blastomogenic effect of lubricating oils
Gig. Tr. Prof. Zabol., 10, 50–51 (1971)

Weatherby, J.H., Williams, G.Z.
Studies on the toxicity of diethylene glycol, elixir of sulfanilamide-Massengill and a synthetic elixir
J. Amer. Pharm. Ass., 28, 12–17 (1939)

Wegener, H.
Über die Fortpflanzungsfähigkeit der Ratte nach Einwirkung von

Diäthylenglykol
Arch. exper. Path. Pharmacol., 220, 414–417 (1953)

Weil, C.S., Carpenter, C.P., Smyth, H.F., Jr.
Urinary bladder response to diethylene glycol
Arch. Environ. Health, 11, 569–581 (1965)

Weil, C.S., Carpenter, C.P., Smyth, H.F., Jr.
Urinary bladder calculus and tumour response following either repeated feeding of diethylene glycol or calcium oxalate stone implantation
Ind. Med. Surg., 36, 55–57 (1967)

Wiley, F.H., Hueper, W.C., Bergen, D.S., Blood, F.R.
The formation of oxalic acid from ethylene glycol and related solvents.
J. Industr. Hyg. Toxicol., 20, 269–277 (1938)

Winek, C.L., Shingleton, D.P., Shanor, S.P.
Ethylene and diethylene glycol toxicity
Clin. Toxicol., 13 (2), 297–324 (1978)

Woo, Y., Arcos, I.C., Argus, M.F., Griffin, G., Nishiyama, K.
Metabolism in vivo of dioxane:
identification of p-dioxane-2-one as the major urinary metabolite
Biochem. Pharmacol., 26, 1535–1538 (1977a)

Woo, Y., Arcos, I.C., Argus, M.F., Griffin, G., Nishiyama, K.
Structural identification of p-dioxane-2-one as the major urinary metabolite of p-dioxane
Naunyn-Schmiedeberg's Arch. Pharmacol. 299, 283–287 (1977b)

1,4-Dicyanobutane

1. Summary and assessment

1,4-Dicyanobutane is metabolized in the animal to hydrogen cyanide and thiocyanate and is excreted mainly as thiocyanate in the urine. It has been suggested that it is hydrolysed by alpha-cyanohydrin; other metabolites, in part unidentified, are probably also present.

From the reported studies it is clear that 1,4-dicyanobutane is acutely toxic in animals when administered orally or by inhalation (LD_{50} rat oral between 138 and 483 mg/kg body weight; LC_{50} in the rat on inhalation 1710 mg/m^3 of air for 4 hours). Observed symptoms include tremor, excitability and convulsions.

Undiluted 1,4-dicyanobutane does not irritate the skin of animals, and is only mildly irritating to the eye. It is absorbed through the skin.

Repeated administration, orally, dermally, subcutaneously or by inhalation, to animals at relatively low doses (10–25 µl/kg) leads chiefly to convulsions, signs of paralysis, loss of body weight and an abnormal blood picture. These changes are reversible. Histological investigation in rats following inhalation exposure (31 to 268 mg/m^3 air, 6 hours daily, 5 days a week for 10 days) revealed no pathological changes.

The only damage observed in a two-year study of rats (the substance being administered in the drinking water) occurred in the adrenal glands. Details of the experimental findings are not available.

In the Salmonella/microsome test, 1,4-dicyanobutane is not mutagenic. The findings of a reproductive study on rats indicate a weak embryotoxic effect. No evidence of a teratogenic effect has been noted. It has been suggested that the observed changes (reduced body weight of foetuses) are due partly to the action of hydrogen cyanide.

In humans, skin contact can result in local irritation of a greater or lesser severity, depending upon the intensity of exposure; a sensitizing effect has not been observed. Administration orally or by inhalation must be expected to result in cyanosis.

1,4-Dicyanobutane

2. Name of substance

2.1 Usual name   1,4-Dicyanobutane
2.2 IUPAC name   1,4-Dicyanobutane
2.3 CAS-No.   111-69-3

3. Synonyms

Common and trade names

Adiponitrile
Adipic acid dinitrile
Tetramethylene dicyanide
Adipyl dinitrile
Butane dinitrile
Hexanedioic acid dinitrile
Tetramethylene cyanide

4. Structural and molecular formulae

4.1 Structural formula

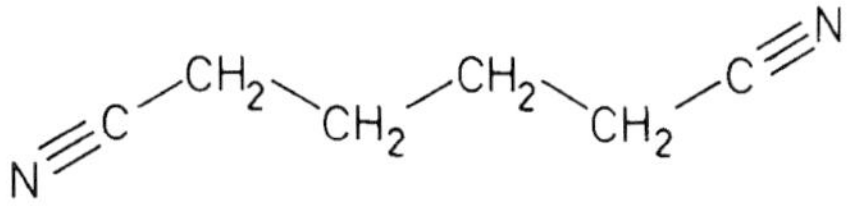

$N\equiv C-CH_2-CH_2-CH_2-CH_2-CH_2-C\equiv N$

4.2 Molecular formula $C_6H_8N_2$

5. Physical and chemical properties

5.1 Molecular mass	108.14 g/mol
5.2 Melting point	2.4° C
5.3 Boiling point	298–300° C from 93° C, decomposition with formation of hydrogen cyanide
5.4 Density	0.965 g/cm^3 (at 20° C)
5.5 Vapour pressure	2.67 hPa (at 119° C)
5.6 Solubility in water	4–5% at 20° C completely miscible at 100° C
5.7 Solubility in organic solvents	soluble in ethanol, chloroform
5.8 Solubility in fat	no information available
5.9 pH-value	–

5.10 Conversion factors	1 ppm $\hat{=}$ 4.505 mg/m^3
	1 mg/m^3 $\hat{=}$ 0.222 ppm
	(Ullmann, 1985; Clayton and Clayton, 1982; Hommel, 1980)

6. Uses

Intermediate product in the manufacture of Nylon 66 (Ullmann, 1985).

7. Results of experiments

7.1 Toxicokinetics and metabolism

Dermal absorption of 1,4-dicyanobutane occurred in the guinea pig in proportion to the quantity applied and the area of the application site, and was independent of the type of solvent and the concentration. This study used dosage levels of 50 to 1000 mg/kg body weight (duration of application 4 hours), application areas of 12.5 and 25 cm^2, the solvents water and alcohol, and concentrations of 1,4-dicyanobutane ranging from 5% to the undiluted product.

On normal undamaged skin (shaved), absorption (0–24 hours) in the guinea pig following 4-hour application was between approx. 1 and 4%, while on the abraded skin it was between 20 and 45% (Ghiringhelli, 1955b).

It is suggested that in the animal 1,4-dicyanobutane might be hydrolysed by alpha-carbon to an alpha-cyanohydrin (alpha-hydroxynitrile) with the release of hydrogen cyanide. However, other metabolites not yet characterised probably also occur (Tanii and Hashimoto, 1985).

1,4-Dicyanobutane is excreted chiefly as thiocyanate in the urine and as hydrogen cyanide in the respiratory air. Thus, following subcutaneous administration, the guinea pig eliminated 79% of 1,4-dicyanobutane as thiocyanate in the urine (0 to 24 hours). For the dog, this value following oral dosing was 50% (time not specified; Svirbely and Floyd, 1964).

In mice dying of acute poisoning (519 mg/kg body weight, orally) a level of 0.71 ± 0.19 µg CN/g tissue was detected in the brain. The animals (n=8) died 83 ± 3 minutes after dosing (Tanii and Hashimoto, 1985). In the case of a guinea pig dying prematurely (time not specified) an average of 0.68 mg% hydrogen cyanide was found in cardiovascular blood samples (Ghiringhelli, 1955a).

The Michaelis constant for the release of CN from 1,4-dicyanobutane was 0.625 mM for mouse liver microsomes (Tanii and Hashimoto, 1985).

7.2 Acute and subacute toxicity

Following a single administration, the LD_{50} values obtained are presented in table 1 (see p. 253 ff.)

The following symptoms of poisoning were seen following oral or intraperitoneal administration:

Mouse   Excitability, convulsions (BASF, 1959)
Rat   Salivation, reduced coordination of movements and also reduced activity, tremor, dose-dependent loss of body weight (Dashiell and Kennedy, 1984).
Rabbit, cat   Shortness of breath, excitability, convulsions, lying on the side, and in some cases cyanosis. The cats also vomited, so that the LD_{50} values quoted above are in actual fact probably even lower (Zeller et al., 1969).

In dermal studies, 1,4-dicyanobutane was applied once, for 1 or 4 hours, to the shaved abdominal skin of Sprague-Dawley rats (5 animals/group). For this purpose the animals were fastened in a tank with 2 ml of the substance. The application area corresponded to approx. 10% of the body surface. Following application, the skin was washed with Polyethylene glycol (PEG) 400 and dried with cellulose. The clinical picture resulting from poisoning was marked by slight convulsions. None of the animals died. Neither systemic nor local changes were in evidence on the macroscopic level (Zeller et al., 1969).

0.5 or 1.0 ml 1,4-dicyanobutane/kg body weight (0.7 or 19.3 mg/kg) was applied to the shaved backs of rabbits (area of application approx. 50 cm^2) and distributed evenly. To prevent evaporation the treated area was covered with parchment paper. The animals were held in position for the duration of the experiment (up to 24 hours). Rabbits survived a dose of 0.5 ml/kg, whilst a level of 1.0 ml/kg led to death (no further details; BASF, 1959).

In the guinea pig an LD_{50} value of 50 mg/kg body weight was obtained with a single subcutaneous injection (no further details; Ghiringhelli, 1956).

In male Charles-River rats an LC_{50} value of 1710 mg/m^3 air (380 ppm) was established for a 4-hour exposure. The main symptoms of poisoning were lethargy, salivation, impeded respiration, convulsions and weight loss. Death occurred within 4 days of exposure (see also Table 1; Smith and Kennedy, 1982).

In a further study with Charles-River rats, the inhalation-risk test was carried out. For 6 animals the exposure period that was survived by all animals during a subsequent observation period of 7 days was established as 8 hours. No symptoms of poisoning were noted (BASF, 1959; Zeller et al., 1969).

In the rabbit, repeated oral administration (25 µl/kg body weight given 4–6 times or 10 µl/kg body weight given 16–36 times (24.1 mg/kg body weight or 9.7 mg/kg body weight) was followed by toxic symptoms of varying intensity, consisting in some animals only of atony and in others of very severe convulsions. All the rabbits showed weight loss, considerably in some cases; urine, blood picture and liver function were unaffected (no further details; Zeller et al., 1969).

In cats, oral dosing with 1,4-dicyanobutane 9 to 17 times at a level of 25 µl/kg body weight (24.1 mg/kg body weight), caused salivation, vomiting, dyspnoea, atony and tonic-clonic convulsions, and in some animals also ataxia or pareses of the hind extremities. These toxic symptoms were only slowly reversible in some cases, so that it became necessary to insert relatively long intervals between treatments (no further details). Following these intervals the same symptoms reappeared; in some animals they seemed to intensify with an increasing number of doses. Following administration, the exhalation air smelt of prussic acid. All the animals showed a considerable decrease in body weight and a marked drop in haemoglobin and erythrocytes. No liver or kidney damage was detected. Dissection revealed no pathological changes in the internal organs on the macroscopic level (no further details; Zeller et al., 1969).

1,4-Dicyanobutane was applied to the shaved skin of 9 guinea pigs (application area 3 cm diameter) every 2nd day for 1 month (no further details on dosage level). A control group was not included. The toxic symptoms were marked by a slightly reduced body weight, reduced calcium level in the blood, hypochromic haemolytic anaemia, and also leucopenia and lymphomonocytosis. An increased erythropoesis was found in the bone marrow. Histologically, the following results were recorded: hyperaemia of the internal organs, fatty degeneration of the liver, enlargement of the epithelial cells of the convoluted tubules, and haemosiderosis of the spleen, as well as hyperaemia, atelectasis and emphysema of the lungs (Ghiringhelli, 1955a).

1,4-Dicyanobutane was administered subcutaneously to 8 guinea pigs at dosage levels of 3, 10, 20 and 30 mg/kg body weight/day on 6 days a week for a period of 40 to 70 days (no further details). This corresponds to 35 to 60 treatment days. A control group was not included. A haematological examination gave no evidence of any effect on these parameters (Ceresa, 1948).

In an investigation of subacute inhalation toxicity, 10 male rats were exposed to 31, 95 or 268 mg 1,4-dicyanobutane/m^3 air (6 hours/day, 5 days/week, 10 days of treatment, total duration of experiment 12 days). 10 further rats served as a control group. 5 rats were killed at the end of the 10th day of treatment, the remaining 5 rats at the end of a subsequent 14-day observation period. During exposure, slightly irregular respiration, slight salivation and reduced reaction to noises occurred at all dosage levels. At the high level a slight reduction of body weight gain occurred during the first 5 days of treatment. The haematological and clinical-chemical investigation at the end of the period of treatment showed a significant ($p \leq 0.05$) reduction in the numbers of erythrocytes and leucocytes and in the haemoglobin value at the high dose; significant increases were seen in the blood-glucose level, the urine volume and the number of animals with raised urine-sugar values. The urea values were significantly ($p \leq 0.05$) below the control values from 31 mg/m^3, as were the number of eosinophils from 95 mg/m^3. All parameters had returned to normal at the end of the subsequent 14-day period of observation. A histological examination was carried out on all animals for the following tissues: eyes, caecum, colon, duodenum, brain, ear, skin, heart, testes, bone marrow, liver, lungs, stomach, mediastinum, spleen, epididymis, adrenals, kidneys, pancreas, thyroid and trachea. No pathological changes could be detected (Smith and Kennedy, 1982).

7.3 Skin and mucous membrane effects

The acute irritancy was investigated on the shaved dorsal skin of white rabbits in the so-called rag test, in which the undiluted substance was in contact with the skin once for 1, 5 or 15 minutes or for 20 hours. Following the 1-, 5- and 15-minute applications the reacted area was washed first with undiluted PEG 400 and then with a 50% aqueous PEG 400 solution; in contrast, following 20-hour exposure the skin remained untreated. The result was monitored on removal of the dressing and after 1, 3 and 8 days. No changes were evident (BASF, 1959).

To test the acute irritant effect on the eye, 1 drop (approx. 50 µl) of 1,4-dicyanobutane was instilled into the conjunctival sac of the rabbit eye and carefully distributed over the eye surface. 1,4-Dicyanobutane caused a slight reddening and a film-like opacity of the cornea. The observed effects had completely subsided after 8 days (Zeller et al., 1969).

7.4 Sensitization

No information available.

7.5 Subchronic-chronic toxicity

Male and female Carworth-Farm Wistar rats received 1,4-dicyanobutane in the drinking water daily, for 2 years, at concentrations of 0.5, 5.0 and 50 ppm (no further details). This corresponds to 0.05, 0.5 and 5 mg/kg body weight/day, assuming a daily water consumption of 100 ml/kg body weight. Water consumption, body weight gain, the weight of the spleen, liver and kidneys and the haematological parameters remained unaffected. Degeneration of the adrenals was observed in the females at all dosage levels and in the male animals at the high dose (no further details; Svirbely and Floyd, 1964).

7.6 Genotoxicity

In vitro. 1,4-Dicyanobutane was tested on Salmonella typhimurium strains TA 1535, TA 1537, TA 1538 and TA 98 in the Ames test for its mutagenic effect in vitro. The experiments were carried out both with S9-mix (metabolising system from rat liver homogenate) and also without S9-mix using concentrations of up to 10,000 µg/plate. In no test variant were there any indications of a mutagenic effect (DHEW, 1978).

In vivo. No information available.

7.7 Carcinogenicity

No information available.

7.8 Reproductive toxicity

From the 6th to the 19th day of gestation, rats (Charles River) received 1,4-dicyanobutane (degree of purity >98%) at dosage levels of 20, 40 and 80 mg/kg body weight/day, administered by stomach tube. 25 animals were used per group, including a control group. The high dosage level was determined in a preliminary experiment as being the maximum tolerated dose. The animals were killed on the 20th day of gestation, and half the foetuses of a litter were examined for skeletal changes, the remaining ones for changes in the organs. One animal died at the middle dosage level, 2 at the high level. The body weight and behaviour of the mothers remained unaffected. The foetal body weight was significantly ($p \leq 0.05$) reduced at the highest dose. The authors suggest that this effect could result from possible production of hydrogen cyanide. No variations, retardations or malformations occurred (Johanssen and Levinskas, 1986).

7.9 Effects on the immune system

No information available.

7.10 Other effects

No information available.

8. Experience in humans

Zeller et al. (1969) reported on 7 accidents at work in which damage to health occurred following skin contact with 1,4-dicyanobutane. The effects on the skin, chiefly burning and reddening, were generally seen 5 to 15 minutes after the wetting of the skin. General disturbances resulting from absorption were not observed; furthermore, hospital treatment was only required in one of the 7 cases. In this patient, 1,4-dicyanobutane was splashed over the right foot and large quantities got into the shoe from above. Severe blistering and sheetwise peeling of the skin subsequently occurred, with secondary infection, so that surgical treatment was required lasting several months.

No sensitizing effect has yet been established (BASF, 1988).

The drinking of several milliliters of 1,4-dicyanobutane ("valutabile a qualche cc.") led, 20 minutes later, to general weakness, headaches, dizziness and vomiting. The sense of balance was impaired, skin and mucosa were cyanotic and respiration and pulse rates were accelerated. In addition, there was hypotonia of the extremities. Following gastric lavage and treatment with grape sugar and sodium thiosulphate, the patient recovered completely (Ghiringhelli, 1955a).

Table 1. Acute toxicity of 1,4-dicyanobutane

Animal species	Number of animals/dose	Sex	Route of exposure	Subsequent observation period (days)	No effect level (mg/kg bw)	LD$_{50}$/LC$_{50}$ (95% confidence interval); mg/kg body weight
Mouse	4	Male	oral	–	–	1.592 mmol/kg bw $\hat{=}$ 172 mg/kg bw (1.261–2.013 mmol/kg bw $\hat{=}$ 136–218 mg/kg bw)
Mouse	4	–	i.p.	7	–	ca 40
Mouse (NMRI)	–	–	i.p.	8	–	ca 48
Mouse	–	–	i.p.	1	–	ca 68
Rat (Charles-River)	10	Male	oral	14	100–130[a]	138 (122–156)
Rat (Charles-River)	10	Male	oral	14	<250[b]	301
Rat (Sprague-Dawley)	–	–	oral	8	–	ca 290
Rat	–	–	oral	–	–	300
Rat	–	–	oral	1	–	ca 483

Table 1 (continued)

Animal species	Number of animals/dose	Sex	Route of exposure	Subsequent observation period (days)	No effect level mg/kg bw LD_0	LD_{50}/LC_{50}* (95% confidence interval) mg/kg body weight
Rat	–	–	s.c.	–	–	200
Rat (Charles-River)	10	Male	inhalation (4 hour)	14	36–71 mg/m^3 air	1.71 mg/m^3 air (0.87–2.66 mg/m^3 air)
Guinea pig	Total 18	–	s.c.	–	–	50 mg/kg bw
Rabbit	–	–	oral/i.p.	8	–	25–50 µl/kg bw $\hat{=}$ 24–48 mg/kg bw
Cat	–	–	oral/i.p.	8	–	25–50 µl/kg bw $\hat{=}$ 24–48 mg/kg bw

[a] Animals were weaned 24 hours prior to application
[b] Not weaned
– No further details
* mg/m^3 air

References

BASF, AG
Unpublished report (1959)

BASF, AG
Communication from the department of industrial medicine and health protection to the Employment Accident Insurance Fund of the Chemical Industry (1988)

Ceresa, C.
Ricerche sperimentali sull'intossicazione da nitrile adipico: NC-(CH2)4-CN
Med. Lavoro, 39, 274–281 (1948)

Clayton, G.D., Clayton, F.E.
Patty's Industrial Hygiene and Toxicology
Volume 2C, 4880-4900 (1982)
John Wiley and Sons, New York

Dashiell, O.L., Kennedy, G.L., Jr
The effects of fasting on the acute oral toxicity of nine chemicals in the rat
J. Appl. Toxicol., 4, 320-325 (1984)

DHEW (Department of Health, Education and Welfare)
Criteria for a Recommended Standard: Occupational exposure to nitriles
U.S. Department of Health, Education and Welfare, Public Health Service, Center for Disease Control, National Institute for Occupational Safety and Health, Rockville, M.D.
Report PB81-225534 (1978)

Ghiringhelli, L.
Tossicita' del nitrile adipico
Nota I: Avvelenamento acuto e meccanismo d'azione
Med. Lavoro, 46, 221–228 (1955a)

Ghiringhelli, L.
Tossicita' del nitrile adipico
Nota II: studio dell'influenza esercitata sulla crasi ematica da una continua somministrazione e possibilita' di assorbimento del composto attraverso la cute
Med. Lavoro, 46, 229–235 (1955b)

Ghiringhelli, L.
Studio comparativo sulla tossicata' di alcuni nitrili e di alcune amidi
Med.Lavoro, 47, 192–199 (1956)

Hommel, G.
Handbuch der gefährlichen Güter, Merkblatt 7
Springer, Berlin (1980)

Johanssen, F.R., Levinskas, G.J., Berteau, P.E., Rodwell, D.E.
Evaluation of the teratogenic potential of three aliphatic nitriles in the rat
Fundam. Appl. Toxicol., 7, 33–40 (1986)

NIOSH (National Institute for Occupational Safety and Health)
Morbidity and Mortality – Weekly Reports
34 (IS), 22S–23S (1985)

Plzak, V., Doull, J.
A further survey of compounds for radiation protection
Chicago Univ. Report AD-691–490 (1969)

Sax, N.I. (ed.)
Dangerous Properties of Industrial Materials
5th Edition, p. 344
van Nostrand Reinhold Company (1979)

Smith, L.W., Kennedy, G.L. Jr
Inhalation toxicity of adiponitrile in rats
Toxicol. Appl. Pharmacol., 65, 257–263 (1982)

Svirbely, J.L., Floyd, E.P.
Toxicologic studies of acrylonitrile, adiponitrile and beta,beta-oxydipropionitrile – III. Chronic studies
USDHEW, Robert A. Taft Sanitary Engineering Center (1964)
Report not seen, information from DHEW

Tanii, H., Hashimoto, K.
Structure – acute toxicity relationship of dinitriles in mice
Arch. Toxicol., 57, 88–93 (1985)

Ullmanns Encyclopedia of Industrial Chemistry
5th edition, Vol. A1, 273–278 (1985)
VCH Verlagsgesellschaft mbH, Weinheim

Zeller, H., Hofmann, H.T., Thiess, A.M., Hey, W.
Zur Toxizität der Nitrile (Tierexperimentelle Untersuchungsergebnisse und werksärztliche Erfahrungen in 15 Jahren)
Zbl. Arbeitsmed., 19, 225–238 (1969)

Dimethyl terephthalate

1. **Summary and assessment**

Dimethyl terephthalate is rapidly and almost completely absorbed from the gastro-intestinal tract. It can also be shown to be absorbed following dermal application or instillation in the conjunctival sac or respiratory tract. Following oral administration, dimethyl terephthalate is almost completely eliminated within 48 hours through the kidneys (up to 80%) and faeces (approx. 10–15%). Elimination is also mainly through the kidneys following other routes of administration. Urine analysis of orally-treated rats reveals terephthalic acid in particular and traces of monomethyl terephthalate, but no dimethyl terephthalate. In mice, on the other hand, monomethyl terephthalate is the main metabolite following oral administration. This suggests that rats and mice hydrolyse dimethyl terephthalate to differing degrees. An accumulation of dimethyl terephthalate in the various tissues of the body does not occur.

Dimethyl terephthalate is slightly toxic (rat oral, LD_{50} >4.6 g/kg or 6.5 g/kg). Symptoms of poisoning include weakness, tremor and ataxia. The available studies give no clear indication of local skin or mucosal irritation or of a skin-sensitizing effect; however, these studies do not conform to current standards. The studies which report irritation to the skin, eye and mucosa of the respiratory tract cannot be assessed due to lack of detailed data.

Dietary administration of high doses of dimethyl terephthalate in the feed to young male and female rats (>1.5% in the feed, $\hat{=}$ >1.5 g/kg/day) for two weeks leads to loss of weight and the formation of bladder stones, the main components of which are terephthalic acid, calcium and proteins. A critical saturation of the urine with terephthalic acid and calcium is probably chiefly responsible for the formation of these stones in the bladder. The bladder stones cause hyperplasia of the bladder epithelium. Lower concentrations of dimethylterephthalate in the feed (1% and less) over 96 days do not give rise to any clinical symptoms in rats apart from a reduction in weight, and there are no effects on the haematological or clinical chemistry parameters or on the histology of the inner organs. No bladder stones are formed.

Inhalation of dimethyl terephthalate dust 58 times is also tolerated (4 hours daily, 5 times a week, concentration 86.4 mg/m^3 (36% of the particles being taken up by the lungs), approx. 4 mg/kg/day, experimental parameters the same as for oral administration). Toxic effects on various organ systems are reported in some studies, but no assessment is possible due to a lack of experimental data.

Repeated intraperitoneal injections of 15% dimethyl terephthalate (suspended in oil) once a week for 107 days caused no damage to rats (no further details given).

The reported toxic effects to piglets (gastro-enteritis, nausea, cough, agitation) of repeated oral administration of dimethyl terephthalate, or of inhalation exposure to the dust, is difficult to assess due to incomplete documentation of the results. The same is true of a study which demonstrated the appearance of antibodies against dimethyl terephthalate, following its repeated administration to rats.

In the Salmonella microsome test (TA 1535, TA 1537, TA 100, TA 98) no increase is found in the number of revertants at concentrations up to 10,000 µg/plate with or without S9-mix. On the other hand, in the micronucleus test, a dose-dependent increase in micronuclei occurs following i.p. administration of dimethyl terephthalate, and chromosome aberrations in the germ cells have been reported in Drosophila.

In a 2-year NCI feeding study with F-344 rats and B6C3F1 mice, there is no increased incidence of tumours attributable to dimethyl terephthalate administration; dimethyl terephthalate is not considered to be carcinogenic.

Dimethyl terephthalate does not affect fertility, nor does it produce any embryotoxic or teratogenic effects when administered orally or on direct application to the respiratory tract.

Dimethyl terephthalate does not produce skin irritation in humans on repeated application. There are no other available reports of the effects on humans.

2. Name of substance

2.1	Usual name	Dimethyl terephthalate
2.2	IUPAC name	1,4-benzene dicarbonic acid dimethyl ester
2.3	CAS-No.	120-61-6

3. Synonyms

Common and trade names
DMT
Dimethyl-p-phthalate
Dimethyl-1,4-benzene dicarboxylate
1,4-Benzene dicarboxylic acid dimethyl ester
Methyl-4-carbomethoxy-benzoate
Terephthalic acid dimethyl ester

4. Structural and molecular formulae

4.1 Structural formula

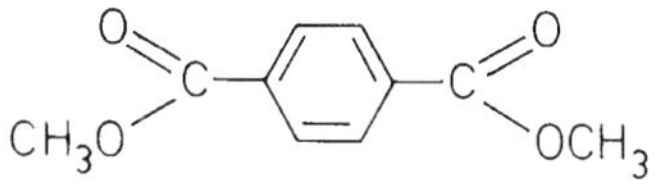

4.2 Molecular formula $C_{10}H_{10}O_4$

5. Physical and chemical properties

5.1 Molecular mass 194.18 g/mol
5.2 Melting point 140.64° C
5.3 Boiling point 280° C
5.4 Density 1.07 g/cm^3 (at 20° C)
5.5 Vapour pressure 11.6 hPa (at 140° C)
5.6 Solubility in water 0.5 g/l (at 20° C)
5.7 Solubility in organic solvents Benzene (2.0 g/100 g at 25° C)
Chloroform (10.0 g/100 g at 25° C)
Dioxane (7.5 g/100 g at 25° C)
Ethyl acetate (3.5 g/100 g at 25° C)
Methanol (1.0 g/100 g at 25° C)
Carbon tetrachloride (1.5 g/100 g at 25° C)
Toluene (4.3 g/100 g at 25° C)
5.8 Solubility in fat No information available

5.9 pH-value —
5.10 Conversion factors   1 ppm $\hat{=}$ 8.057 mg/m^3
  1 mg/m^3 $\hat{=}$ 0.124 ppm
  (Ullmann, 1982; Moffit et al.
  1975; Hoechst, 1986)

6. Uses

Manufacture of polyester fibres (Ullmann, 1982)

7. Results of experiments

7.1 Toxicokinetics and metabolism

When 5% Dimethylterephthalate (DMT) was administered in the diet of rats for 5 days, the DMT was almost completely absorbed from the intestine and was eliminated via the kidneys. Only a trace of the ester was found in the urine, the remainder having been metabolised to terephthalic acid. About 15% of the unabsorbed ester was recovered in the faeces (no further information; Du Pont,1958).

An investigation into the uptake, distribution and elimination of ^{14}C-DMT following oral, intratracheal and dermal administration to male rats or application to the conjunctival sac of the eye in male rabbits yielded the following results:

Following single or repeated oral administration, ^{14}C-DMT is rapidly absorbed from the stomach. After a single dose of ^{14}C-DMT (20 and 40 mg/animal) 75–80% of the administered radioactivity was eliminated in the urine within 48 hours, less than 10% being eliminated in the faeces. Total elimination (urine and faeces) was 83–85%. On repeated oral administration of the same doses of ^{14}C-DMT (5 times within 10 days), 77–79% of the ^{14}C activity was eliminated in the urine and 14–16% in the faeces (total elimination 91.4–95%) within 24 hours of the last dose. Ten days after the end of the test, less than 0.1% of the administered dose was present in the liver, lungs, heart, kidneys, spleen, adrenals, pancreas, testicles, brain and femur (Moffit et al., 1975).

Following a single intratracheal administration of 5 or 10 mg ^{14}C-DMT/animal, roughly 57% of the radioactivity was detected in the urine and some 3% in the faeces within 48 hours. On repeated administration (5 times within 10 days), 54% of the radioactivity was found in the urine and 2.3% in the faeces. Less than 1% of the radioactivity was found in the lungs and tracheal lymph nodes 24 hours after the last administration, and less than 0.1% could be found in the other organs investigated (same organs as for oral administration; Moffit et al., 1975).

Following a single covered application of ^{14}C-DMT (80 mg in 0.2 ml aqueous Triton X solution) to the shaved skin of the backs of rats, 9.3% of the radioactivity was found in the urine and 1.5% in the faeces. Following repeated administration (the covering was replaced for each new application, 5 times in 10 days) 10% and 2.4% of the radioactivity was found. Less than 1% was present in the other organs (the same as for oral administration). No skin irritation was observed when the covering was removed 10 days after the start of the test (Moffit et al., 1975).

50 mg of ^{14}C-DMT was instilled into the conjunctival sac of the rabbit eye. In one test group this was washed with distilled water after 5 minutes, in a second group washing took place after 24 hours. After 5 minutes' contact, 27% of the radioactivity was found in the urine and 2% in the faeces within 10 days; after 24 hours contact, the equivalent amounts were 35% and 2%. Less than 0.1% was recovered from all the measured organs (the same organs as for oral administration), including the eyes. Following extraction of the urine with a chloroform-methanol mixture, ^{14}C activity was found mainly in the aqueous fraction. According to the authors, DMT is rapidly converted to water-soluble metabolites (no information on the chemical structure of the metabolites; Moffit et al., 1975).

In F-344 rats which had received ^{14}C-DMT orally (no further information) HPLC-investigation of the urine revealed that all of the radioactivity of the DMT had been converted to terephthalic acid. Only traces of monomethyl-^{14}C-terephthalate and no ^{14}C-DMT were found, from which the authors concluded that in F-344 rats DMT is almost completely hydrolysed to terephthalic acid (Heck and Kluwe,1980).

In contrast to rats, when the same method of investigation was applied to B6C3F1 mice it was found that the radioactivity of ^{14}C-DMT was converted to monomethyl-^{14}C-terephthalate. Rats therefore appear to hydrolyse DMT to terephthalic acid better than mice (Heck and Tyl, 1985).

No indications of dermal absorption were found when 5 g/kg of DMT (aqueous suspension) was applied to the skin (covered) of guinea pigs for 24 hours (no further information; Eastman Kodak, 1976).

7.2 Acute and subacute toxicity

An oral LD_{50} of >6590 mg/kg and an intraperitoneal toxicity of 3900 mg/kg were determined for Long-Evans rats. With both methods of administration, the treated rats in all the dose groups showed general weakness. At the two highest doses (5020 mg/kg i.p., 6590 mg/kg orally) trembling occurred followed by apathy and ataxia. On dissection and histological examination (no further information) no pathological findings were reported apart from irritation in the peritoneum with foreign body reaction in the case of i.p. administration (Krasavage et al., 1973).

In a further investigation in rats, an oral LD_{50} of >4.6 g/kg and an intraperitoneal LD_{50} of 3.65 g/kg were determined (Massmann, 1966). According to tests by Eastman Kodak (unpublished study) on mice and rats, the oral LD_{50} was >3.2 g/kg and intraperitoneal LD_{50} values of 3.2 and 1.6 to 3.2 g/kg were reported (Eastman Kodak, 1976).

In a further unpublished study, LD_{50} values of >10 g/kg and 3.0 g/kg were determined following oral administration to rats and intraperitoneal injection in mice (BASF, 1961).

The inhalation of DMT at concentrations of 1000–6000 mg/m^3 (vapours and aerosol, ratio not known) for 30 to 60 minutes resulted in irritation with hyperaemia of the mucous membranes, a decrease in erythrocytes and haemoglobin and an increase in the neurological irritation threshold (Sanina et al., 1963).

The dermal LD_{50} (guinea pigs) was >5 g/kg (Eastman Kodak, 1976).

When high doses of DMT were administered in the diet (at concentrations of 0.5, 1.0, 1.5, 2.0 and 3.0% in the food) to young weaned male and female Fischer 344 rats (28 days old, feed intake roughly 500 to 3000 mg/kg/day) for 2 weeks, there was a minimal increase in body weight and dose-dependent bladder stone formation, which at the maximum dose occurred in all male and in about 50% of female rats. No bladder stones developed at concentrations of less than 1.5%. Macroscopic changes detectable when stones were present included thickening of the bladder wall and occasional hydronephrosis. Haematuria was frequently observed in the higher dose group. Analysis of the bladder stones showed the main components to be terephthalic acid, calcium and protein. DMT induced dose-dependent acidosis in the urine and hyperplasia of the bladder epithelium. Over-saturation of the urine with terephthalic acid and calcium was considered to be the cause of bladder stone formation following DMT administration. This is probably also responsible for the occurrence of hyperplasia of the bladder epithelium (Chin et al., 1981; Heck and Tyl, 1985; Heck, 1981).

In a 28-day feeding test in rats (strain and sex not specified), a concentration of 5% caused loss of weight, reduced food intake and higher mortality (no data; Fassett, 1963, 1981).

Rats (number and strain not specified) received 5000 mg/kg orally by stomach tube, 5 times weekly for 2 weeks. DMT produced increasing loss of weight, and 5/6 rats died on the 11th day and showed signs of starvation on dissection(no further details; Du Pont, 1958).

Piglets (2- to 4-months old, 5 animals/group) were administered DMT in the diet at doses of 26 and 260 mg/kg and a higher dose (not specifically defined) for 30 days. The low dose was based on the maximum permissible concentration of 0.1 mg/m^3 in respirable air which may be released into the atmosphere by a factory; each piglet therefore received 22 to 30 mg DMT/kg body weight. The piglets in the two high dose groups became restless, scratched themselves, developed diarrhoea and showed a lower weight gain than the controls. Increased variations occurred in leucocyte count and immunoglobulins, ruling out an immune reaction. On dissection, gastro-enteritis, inflammation of the bronchi and lungs and localized changes in the liver and kidneys were found. When the meat of the animals treated with high doses was eaten it was found to have a specific after-taste, and the cooked meat was a yellowish grey colour while that of the untreated controls was light grey with a reddish hue. Thin layer chromatography revealed traces of DMT in the stomach, intestine, muscular tissue, liver, kidneys and skin in the the two high dose groups (no specific details; Zhuk and Karput, 1986).

In a second test, piglets (2- to 4-months old, 5 animals/group) were exposed to DMT concentrations of 0.1, 10 and 100 mg/m^3 for 24 hours (no specific test details). Here again, restlessness and coughing were observed at the two higher concentrations. As for oral administration, similar changes were found in the blood counts and pathological investigations of the organs of these animals, and when their meat was tasted. Traces of DMT were again found (by thin layer chromatography) in the lungs, muscles, kidneys, liver and skin. Evaluation of this work is not possible because detailed information on the methods and results is lacking (Zhuk and Karput, 1986).

7.3 Skin and mucous membrane effects

A 50% suspension in distilled water (semi-occluded, shaved skin of white rabbits, exposure time 20 hours) produced no irritation (BASF, 1961).

Single and repeated (4 times) open applications of 0.5 ml of a 1% DMT solution in a mixture of acetone/dioxane/guinea pig fat (7:2:1) caused no irritation on guinea pig skin (Krasavage et al., 1973).

Covered application of 80 mg DMT in 0.2 ml of distilled water to rabbit skin once and 5 times in 10 days (exposure time not given) produced no irritation (Moffit et al., 1975).

5 g/kg of DMT (moistened with distilled water) applied to guinea-pig skin for 24 hours (covered) produced only mild irritation (Eastman Kodak, 1976).

When DMT was applied repeatedly to rabbit skin as a 5% preparation (accurate test data lacking), pigmentation and slight skin irritation developed at the site of application (Sanina and Kochetkova, 1963).

Powdered DMT (0.05 g) applied to the conjunctival sac of the rabbit eye caused only slight reddening which persisted for a few hours (BASF, 1961).

A 15% solution of DMT in oil (no quantitative data) produced no irritant effects on the mucosa of the guinea-pig eye (Massmann, 1966).

Two drops of a suspension caused irritation of the mucosa of the eye (Sanina and Kochetkova, 1963).

The afore-mentioned investigations do not comply with current test requirements.

7.4 Sensitization

Repeated applications (four times) of 0.5 ml 1% DMT in a mixture of acetone/dioxane/guinea-pig fat (7:2:1) caused no skin sensitization in guinea-pigs when an untreated portion of skin was challenged with 0.5 ml of the same preparation. In addition, 10 guinea-pigs previously treated in the same way were challenged using the "foot pad" technique, and again there were no indications of skin sensitization (Krasavage et al., 1973).

7.5 Subchronic and chronic toxicity

Three groups of 30 newly-weaned male rats (Long-Evans strain) received concentrations of 0.25%, 0.5% and 1% DMT in the food for a period of 96 days. In addition, there was a control group of 30 animals. Haematological tests (haematocrit, haemoglobin, leucocytes, differential blood count) and clinical chemical investigations (SGOT, AP, ornithine carbamyl transferase, urea, glucose, serum proteins) were performed on the 55th and 90th days. After 96 days, 10 animals in each group were sacrificed and examined. A further 20 animals/group were observed over their entire lifetime after the administration of DMT had been discontinued. The only observed effect of DMT was in the highest dosage group (1% DMT in the diet) where weight gain was significantly lower ($p < 0.05$) than in the control animals. Food intake in all the treated groups was identical to that of the control animals. The haematological and clinical/chemical parameters showed no pathological changes. Average body weight and relative and absolute kidney weights did not differ from controls. Histological examination of the internal organs (no precise details) did not reveal any abnormal findings. The surviving animals were clinically normal 1 year after the end of DMT treatment; haematology, enzyme determinations and histology for some sacrificed animals were normal (Krasavage et al., 1973).

28-Day-old male and female Wistar rats (initial weight 56 and 55 g) received DMT at concentrations of 0.5, 1.6 and 3.0% in the food for 13 weeks. DMT-induced bladder stones occurred in 12/16 males and 6/16 females in the high-dose group, in 1/19 males in the medium-dose group (1.6% DMT in the food) and in 2/19 males in the low-dose group. Mild hyperplasia of the bladder epithelium was found in 11/16 males and 7/16 females in the high-dose group (3%), and there was a close correlation between the existence of bladder stones and the development of hyperplasia of the bladder epithelium (Vogin, 1972).

In dogs given oral doses of 100 mg/kg/day over 40 days or 200 mg/kg/day over 62 days, increased irritability was seen in 2/4 animals and there was a downwards trend in systolic and diastolic blood pressure in 3/4 animals. No pathological or histological changes were seen (no further information; Du Pont, undated).

Inhalation studies were performed with 30 rats/group (Long-Evans rats) at DMT concentrations of 86.4 mg/m^3, 16.5 mg/m^3 and 0 mg/m^3 (controls). The mean aerodynamic particle size was 6.6 μm; 36% of the particles could penetrate the lungs. With a 4-hour inhalation period, the total daily intake was calculated to be 4.0 and 0.7 mg/kg body weight. Inhalation exposure was repeated 5 days/week for a total of 58 times. During exposure the rats in the high-dose group rubbed their noses, groomed themselves and blinked. The animals in the low-dose group behaved normally. One day after the last exposure, 10 animals/group were sacrificed and examined. 20 further animals were observed over their entire lifetime. The same haematological, clinical-chemical and histopathological investigation in the animals used were performed as in the afore-mentioned feeding test. There were no striking changes in the measured parameters in any group that could be attributed to the inhalation of DMT. Similarly, there were no abnormal findings one year after the end of treatment (Krasavage et al., 1973).

In another inhalation study, adult Sprague-Dawley rats and male Hartley guinea-pigs were exposed to a DMT dust concentration of 15 mg/m^3 for 6 hours daily, 5 times per week for 6 months. The concentration of the respirable dust was 5 mg/m^3. Body and organ weights (lung, liver, kidneys, spleen), routine clinical-chemical investigations (SMA 12 screening) and urine analysis were unchanged. Macroscopical and histological findings (no information on the organs examined) were normal (Lewis et al., 1982).

When 1 to 4 mg DMT/m^3 (vapour and aerosol) were inhaled for 2 hours/day for 5 months, the exposed rats manifested nervous system effects, low blood pressure, slight anaemia and reticulocytosis. Chronic inflammation of the respiratory tract was noted on histopathological examination (no precise details; Sanina and Kochetkova, 1963).

When groups of rats were exposed to DMT vapour and aerosol at concentrations of 40 to 70 mg/m^3 for 2 hours/day over 5 months, 30% of the animals died after 2.5 to 3 months. On dissection, there was inflammation of the respiratory tract (rhinitis, tracheitis) and emphysema of the lung, together with dystrophic changes in the liver and kidneys (no precise information; Sanina and Kochetkova, 1963).

Ten female rats (strain not specified) received an intraperitoneal injection of a 15% suspension of DMT in olive oil every 7th day over a period of 109 days (some 1 to 1.7 ml/animal, total dose 3.5 g/animal). All animals appeared normal during this period. One rat died after 56 days. After the test all the animals were dissected, but there were no pathological findings. Weight gain in the treated animals slowed down after week 6 in comparison with the controls receiving olive oil alone (Massmann, 1966).

Wistar rats were exposed to DMT concentrations of 0.08, 0.4 and 1 mg/m^3 for 24 hours daily for 3 months. In addition, 37 Wistar rats received DMT orally at doses of 0.075, 7.5 and 75 mg/kg for 30 days. In these animals, degranulation of the basophilic leucocytes, platelet-formation (after Jerne), complement-fixation and lymphocyte transformation were investigated. An aqueous solution of DMT and a 25% aqueous salt extract of the lung tissue of treated animals were used as antigens. Both on inhalation exposure and on oral administration, the adminstered concentrations and doses resulted in increased degranulation of the basophils, increased platelet-formation, a positive complement-fixation reaction and lymphocyte transformation (all partially dose-dependent). The gamma-globulin fraction from the blood serum of the inhalation test was administered to pregnant rats. Following this, increased foetal death, increased bleeding and an increased number of rib deformities were described (no precise details). From the investigation it was concluded that DMT has a sensitising effect and that the antibodies present in blood serum have an embryotoxic effect. However, the methods and results are insufficiently described for the work to be assessed (Vinogradov et al., 1986).

7.6 Genotoxicity

DMT was tested up to the maximum tolerable concentrations in the Salmonella microsome test (Salmonella typhimurium TA 98, TA 100, TA 1535,TA 1537) with and without S9-mix from Aroclor-1254-induced livers of rats and hamsters. Doses of 33 to 333 µg/plate showed no mutagenic effect (Zeiger et al., 1982).

In another Ames test, DMT was negative up to 10.000 µg/plate (Heck and Tyl, 1985).

DMT was likewise negative in an Ames test with TA 100 and TA 98 (no information on dosage; Kozumbo et al., 1982).

DMT was not mutagenic in another study using the Salmonella microsome test with and without metabolic activation (no further data; Du Pont, 1979).

In the Drosophila test, the frequency of dominant lethal mutations (DLM) in males which had been treated by the imago feeding method increased at concentrations of 150 mM in the nutrient medium (DLM: 7.35 ± 0.50, controls 4.30 ± 0.78, $p < 0.001$; Goncharova et al., 1984).

DMT was tested in a micronucleus test on male mice (C57Bl/6jxCBH-F1, 15 mice/group) with a single i.p. injection (in DMSO) at doses of 0.02, 0.25, 0.33, 0.5 and 1.0 mmol/kg (ca. 39, 49, 64, 97 and 194 mg/kg). The number of micronuclei was significantly increased in a dose-dependent way in the test groups. At the highest dose, depression of the bone marrow was observed 24 hours after i.p. injection. Only typically coloured particles in the cell with diameters of between 1/20 and 1/5 of the cell diameter were regarded as micronuclei (Goncharova et al., 1984).

7.7 Carcinogenicity

In a lifetime test (NCI, 1979) 50 male and 50 female Fischer 344 rats and 50 male and 50 female B6C3F1 mice received a diet of 2500 and 5000 ppm DMT for 103 weeks. There followed an additional 2 week observation period. Control groups of male and female rats and mice of equal size were maintained. The animals were some 6 weeks old at the start of the test. All surviving rats were sacrificed at an age of 111 or 112 weeks, all surviving mice at 110 or 111 weeks. The administration of DMT had no effect on body weight gain or on survival time. Clinical observations did not reveal any treatment-related effects. A kidney stone was found in only one female rat. The number and nature of the tumours occurring in male and female rats did not differ significantly from those in controls. Of the male mice, 8/49 animals developed lung tumours (alveolar and broncheolar adenomas or carcinomas) in the low-dose group, 13/49 in the high-dose group. In the control group for this study the number was 1/49 ($p < 0.01$). The authors considered the indidence of tumours in the control group to be extraordinarily low, as other control groups with the same strain kept in the same area showed a considerably higher number of corresponding lung tumours (5/49, 6/46 and 9/49). When the number of tumours occurring in the treated groups was compared with those in the latter control groups, no significant differences were found. Lymphomas and hepatomas occurred in treated female mice, but at a lower incidence than in the control group. It was therefore concluded that under the conditions of this test DMT was not carcinogenic to rats and mice. It was also stated that the absence of general toxic manifestations indicated that the doses used were not sufficiently high to achieve maximum test sensitivity (NCI, 1979; Rall, 1981).

7.8 Reproductive toxicity

Male rats (Long-Evans strain, 20 animals/group) received a diet containing DMT concentrations of 0.25%, 0.5% and 1% for a period of 115 days. Virgin females received this DMT concentration in the food for 6 days prior to mating. Mating took place 1:1 over 7 days. DMT administration continued over this period. After this, DMT continued to be administered in the diet of the females until the end of the lactation period (21 days post-partum). The males were observed on a normal diet. All the male rats showed normal sexual behaviour, 95 to 100% of the copulations resulted in pregnancy in the females, and litter size was normal. There was no increase in the mortality of the offspring at the time of birth or at weaning. A dose-dependent reduction in the average body weight of the young animals in the two high-dose groups was reported up to the time of weaning (p <0.05; Krasavage et al., 1973).

A single dose of 1000 mg DMT/kg body weight was administered by stomach tube to a group of 20 female Wistar rats from the 7th to the 16th day of pregnancy. Controls treated at the same time received the vehicle alone (starch paste). The mother animals were sacrificed on the 21st day of pregnancy and the foetuses were examined morphologically for developmental defects. The investigation showed that repeated oral administration of 1000 mg/kg body weight in the sensitive phase of organogenesis did not result in any adverse effect on the general health of the mothers nor any disturbance of intrauterine development. The foetuses obtained from the treated group developed normally and did not show any treatment-related abnormalities or variations in the external appearance of the inner organs or of the skeleton. There was no indication that the test substance had any teratogenic potential. The no-effect level for DMT for maternal and embryo/foetal toxicity in rats thus lay at 1000 mg/kg body weight (Hoechst, 1986).

In an inhalation test, 30 pregnant rats at the same stage of pregnancy (20 days) were exposed to a DMT concentration of 1 mg/m^3 for 24 hours daily, and according to the author's report toxic symptoms occurred in the mothers. The control group consisted of 17 pregnant rats. All the animals were sacrificed on the 20th day of pregnancy and the foetuses were examined. There were 208 live foetuses in the DMT-treated group. The number of dead embryos in this group was 84 (28.7%), compared with 8 (4.8%) in the controls. The authors considered this increase in embryo mortality to be insignificant because mortalities of 15% or more had been found in other control groups. No deformities were found in the 208 live foetuses in the DMT group. Important details of the test methods and results were not given in this report (Krotov and Chebotar, 1972).

7.9 Effects on the immune system
See section 7.5

7.10 Other effects
No information available

8. Experience in humans

An oily paste containing 80% DMT showed no irritant effects 24 hours after 10 applications to human skin (no further information; Massmann, 1966).

It has also been reported that itching dermatitis can occur following exposure to DMT dust (Eastman Kodak, 1976; without specific details). In addition to this, irritation of the respiratory tract by vapour or dust has been observed (American Industrial Hygiene Association).

References

American Industrial Hygiene Association
Dimethylterephthalate
Workplace environmental exposure level guide (WEEL)

BASF AG
Unpublished investigation (1961)

Chin, T.Y., Tyl, R.W., Popp, J.A., Heck, H.
Chemical urolithiasis
Toxicology and Appl. Pharmacology, 58, 307–321 (1981)

Du Pont, Haskell Laboratory
Unpublished data (undated) MR-0227-001 cited by American Industrial Hygiene Association

Du Pont, Haskell Laboratory
Unpublished data (1958) cited by American Industrial Hygiene Association

Du Pont, Haskell Laboratory
Unpublished data (1979) cited by American Industrial Hygiene Association

Eastman Kodak
Eastman Chemical Products Inc., Kingsport, TN
Dimethyl Terephthalate
Publication No. GN-309A (1976) cited by American Industrial Hygiene Association

Fassett, D.W.
Laboratory of Industrial Medicine, Eastman Kodak Company

Unpublished data cited in F.A. Patty, Industrial Hygiene and Toxicology, 2nd edition, Vol. II (1963)
3rd revised edition, Vol.II (1981)

Goncharova, R.J., Kuzir, T.D., Levina, A.B., Zabrejko, S.P.
Mutagenic activity of dimethyl terephthalate
Proc. Acad. Sci. (B.S.S.R.) XXVIII, 1041–1044 (1984)

Goncharova, R.J., Zabrejko, S., Kozachenko, V.J., Pashin, Y.V.
Mutagenic effects of dimethyl terephthalate on mouse somatic cells in vivo
Mutat. Res., 204, 703–709 (1988)

Heck, H. d'A.
Chemical urolithiasis, 2. Thermodynamic aspects of bladder stone induction by terephthalic acid and dimethylterephthalate in weaning Fischer-344 rats
Fundam. Appl. Toxicol., 1, 299–308 (1981)

Heck, H. d'A., Kluwe, C.L.
Microanalysis of urinary electrolytes and metabolites in rats ingesting dimethylterephthalate
J. Anat. Toxicol., 4, 222–226 (1980)

Heck, H. d'A., Tyl, R.W.
The induction of bladder stones by terephthalic acid, dimethylterephthalate and melamine (2,4,6-Triamino-s-triazine) and its relevance to risk assessment
Regul. Toxicol. and Pharmacol., 5, 294-313 (1985)

Hoechst AG
Dimethyl terephthalate, investigation of embryotoxic action in Wistar rats on oral administration
Unpublished report, no. 86.0859 (1986)
Commissioned by the Employment Accident Insurance Fund of the Chemical Industry

Kozumbo, W.J., Kroll, R., Rubin, R.J.
Assessment of the mutagenicity of phthalate esters
Environ. Health Perspect., 45, 103–109 (1982)

Krasavage, W.J., Yanno, F.J., Terhaar, C.U.
Dimethyl terephthalate (DMT): Acute toxicity, subacute feeding and inhalation studies in male rats
Am. Ind. Hyg. Ass. J., 34, 455–462 (1973)

Krotov, Y.A., Chebotar, N.A.
A study of embryotoxic and teratogenic action of some industrial substances formed during production of dimethylterephthalate
Gig. Tr. Prof. Zabol., 16,40–43 (1972)

Lewis, T.R., Lynch, D.W., Schuler, R.L.
Absence of urinary bladder and kidney toxicity in rats and guinea pigs exposed to inhaled terephthalic acid and dimethylterephthalate
Toxicologist, 2, 7 (Abstr. 25) (1982)

Massmann, W.
Evaluation of the occupational hygiene/toxicology of p-toluic acid methyl ester, dimethylterephthalic and terephthalic acid
Institute of Occupational Medicine, University of Tübingen, 26.2.1966

Moffit, A.E., Clary, J.J., Lewis, T.R., Blanck, M.D., Perone, V.B.
Absorption, distribution and excretion of terephthalic Acid and dimethyl terephthalate
Am. Ind. Hyg. Ass. J., 36, 633–641 (1975)

NCI (National Cancer Institute)
Carcinogenic Testing Program: Bioassay of dimethylterephthalate for possible carcinogenicity
NIH Publication no. 79–1376, 1979

Rall, D.P.
Reevaluation by the National Toxicology Program of technical report NCI-CG-TR-121 entitled bioassay of dimethylterephthalate for possible carcinogenicity
Federal Register, Vol. 46, Nr. 238, December 11, 1981

Sanina, Y.P., Kochetkova, T.A.
Toksikol. Novyky, Prom. Khim. Veshchestv.No. 5, 107–123 (1963) (Chemical Abstract 61, 6250 f) cited by American Industrial hygiene Association

Ullmanns Enzyklopädie der technischen Chemie
4th Editon, Vol. 22, p. 529–533
Verlag Chemie, Weinheim (1982)

Vinogradov, G.I., Vinarskaja, E.J., Antomonov, H.J., Naumenko, G.M., Gonchar, N.M., Leonskaja, G.J.
Hygienic assessment of the allergy-inducing activity of dimethyl-

terephthalate on oral administration and inhalation exposure in the organism
Kiev. N11 Obshch. Kommunal'n Gig., Kiev, USSR
Gig. Sanit., 5, 7–10 (1986)

Vogin, E.E.
Subacute feeding studies (13 weeks) in rats with dimethylterephthalate (DMT), isophthalic acid (IA) and terephthalic acid (TA)
Food and Drug Research Laboratories, Maspeth NY (1972) cited by Heck & Tyl (1985)

Zeiger, E., Haworth, S., Speck, W., Mortelmans, K.
Phthalate ester testing in the National Toxicology Program's environmental mutagenesis test development program
Environ. Health Perspect., 45, 99–101 (1982)

Zhuk, L.L., Karput, J.M.
Influence of dimethylterephthalate (DMT) on pig productivity
Vestsi, Akad, Navuk, BSSR. Ser. Sel'skagaspad, Navuk, 4, 107–112 (1986)

3-Methylbutanol-1

1. **Summary and assessment** _____

In humans, 3-methylbutanol-1 is rapidly metabolized to isovaleraldehyde and isovaleric acid. Following oral or parenteral administration, no or only a very low concentration of 3-methylbutanol-1 is detectable in the blood shortly after application, while no unchanged substance is found in the urine. Small amounts are eliminated as glucuronide. In contrast to ethanol, 3-methylbutanol-1 distributed in aqueous as well as anhydrous tissues.

The acute oral toxicity of 3-methylbutanol-1 (isoamylalcohol) is low (oral LD_{50} for the rat is 1.3–7.8 g/kg). The following toxic symptoms have been observed in acute trials using various modes of administration: apathy, atonia, paralysis of the pelvic extremities, necrosis, irritation of the mucosa and loss of corneal reflexes. Histologically, findings of degenerative alterations in the liver and kidneys are reported.

3-Methylbutanol-1 has a moderately irritating effect on both intact and abraded skin of rabbits in occlusive tests, while testing on the eye of the rabbit shows definite irritation of the mucosa.

In a subchronic study on rats over 17 weeks, 3-methylbutanol-1 administered orally (dosage up to 1000 mg/kg daily) is tolerated with no detectable adverse effects.

No evidence of a teratogenic effect has been reported.

The experimental findings reported in the literature on the carcinogenic effect of 3-methylbutanol-1 following oral and subcutaneous administration to rats are not conclusive, since the presented data do not permit either a qualitative or a quantitative interpretation of the results. A 90-day study to determine the question of a possible carcinogenic effect has been undertaken by the BG Chemie and will, if indicated, be followed by a study of carcinogenicity. Furthermore, an investigation of possible teratogenic effects is being carried out on behalf of the BG Chemie.

In humans, a concentration of >100 ppm of 3-methylbutanol-1 results in irritation of the eyes, the nose and the throat in the majority of test subjects. Oral administration of ethanol and 3-methylbutanol-1 (which is present naturally in alcoholic drinks) leads in the alcoholization phase to a deterioration of performance, which correlates positively with the ethanol concentration in the blood. In the post-alcoholization phase, the presence of 3-methylbutanol-1 results in a considerable increase (depending on its concentration) in the number of errors.

The odour threshold in the air is given as 0.042 ppm.

The MAK value (maximum workplace concentration) for 3-methylbutanol-1 is set at 100 ppm (360 mg/m^3; Deutsche Forschungsgemeinschaft, 1987).

2. Name of substance

2.1 Usual name   3-methylbutanol-1
2.2 IUPAC name   3-methylbutanol-1
2.3 CAS-No.   123-51-3

3. Synonyms

Common and trade names   Isoamyl alcohol
                                         Isopentyl alcohol
                                         Isobutylcarbinol

4. Structural and molecular formulae

4.1 Structural formula

$$CH_3-CH(CH_3)-CH_2-CH_2-OH$$

4.2 Molecular formula $C_5H_{12}O$

5. Physical and chemical properties

5.1 Molecular mass   88.15 g/mol
5.2 Melting point   117.2° C
5.3 Boiling point   132° C
5.4 Density   0.809 g/cm^3 (at 20° C)
5.5 Vapour pressure   3.73 hPa (at 20° C)
5.6 Solubility in water   low (2 g/100 ml at 14° C)

5.7	Solubility in organic solvents	glacial acetic acid, ethanol, ether, chloroform, acetone, petroleum ether, benzene
5.8	Solubility in fat	no information available
5.9	pH-value	–
5.10	Conversion factor	1 ppm $\stackrel{\wedge}{=}$ 3.6 mg/m^3 1 mg/m^3 $\stackrel{\wedge}{=}$ 0.278 ppm (Windholz, 1983; Weast, 1981/1983; Ullmann, 1974)

6. Uses

Solvent for oils, fats, resins, waxes, and in the plastics industry in the spinning of polyacrylonitrile.

7. Results of experiments

7.1 Toxicokinetic and metabolism

Following oral administration to chinchilla rabbits (3 kg) of 3-methylbutanol-1 in aqueous solution at a dose of 25 mmol (2.2 g/animal), 9% of the dose was excreted in the urine within 24 hours, in the form of the glucuronide. The glucuronide was identified as triacethyl-beta-isoamylglucuronide-methylester (no further details were given; Kamil et al., 1953).

Gaillard and Derache (1964, 1965) investigated the rate of elimination of 3-methylbutanol-1 from the blood, after oral administration of 2 g/kg in 20% aqueous emulsion (tests on 5 Wistar rats, average weight 300 g, sex not stated). Blood was taken from the tail vein for analysis after 0, 15, 30, 60, 90, 120, 240 and 480 minutes. Urine was also collected during this period and tested for the amount of 3-methylbutanol-1 excreted. 7 mg/100 ml were found in the blood after 15 minutes. A peak concentration of 17 mg/100 ml occurred after 1 hour. After 8 hours, 1 mg/100 ml could still be detected. No unchanged 3-methylbutanol-1 was excreted in the urine. The level of alcohol was determined according to Widmark. The same authors found no effect of 3-methylbutanol-1 (4 g/kg in 50% aqueous solution, orally administered) on the lipid content of the liver and blood of rats (tests of 8 Wistar rats, 220 g). 17 hours after administration of the alcohol, the lipid content of the liver and the concentration of triglycerides, cholesterol and phospholipids in the blood were unchanged (Gaillard and Derache, 1966).

4 g water/kg containing 600 mg of 3-methylbutanol-1 (2.4 mg of 3-methylbutanol-1/kg) were administered intraperitoneally to 10 albino rats (no other details). Following administration the concentration of 3-methylbutanol-1 and isovaleraldehyde (intermediate metabolite of 3-methylbutanol-1) in the blood of the test animals was determined at 2-hour intervals by gas chromatography. 2 hours after administration of the substance, neither unchanged alcohol nor isovaleraldehyde were detectable in the blood. In contrast, the administration of 3-methylbutanol-1 together with ethanol resulted in detectable blood levels of 3-methylbutanol-1 after 10 hours and isovaleraldehyde after 8 hours (Greenberg, 1970).

5 ml of rat liver homogenate (0.25 g liver/ml) were incubated with 80 µmol of 3-methylbutanol-1 for 30 minutes. In this period, 0.125 mmol of 3-methylbutanol-1/g liver was metabolised; the equivalent amount for ethanol was 0.16 mmol/g liver (Hedlund and Kiessling, 1969).

In the liver perfusion preparation in situ, 5.3 µmol of 3-methylbutanol-1 in 200 ml of perfusion fluid (heparinized human blood diluted 1:1 with physiological NaCl solution) was perfused for 30 minutes (2 ml/minute). Under these test conditions, 0.04 mmol of 3-methylbutanol-1/g liver was metabolised in this period. This was less than for ethanol (approx. 0.05 mmol/g liver; Hedlund and Kiessling, 1969).

Kühnholz et al. (1984) determined the solubility of aliphatic alcohols in the body tissue of rabbits and man in vitro (5 ml of physiological saline solution with a concentration of 10 mg/l of the alcohol were incubated with pulverized body tissue for a minimum of 45 minutes at 37° C). Muscles, brain, lungs, kidneys, fatty tissue, liver, spleen and blood were tested. Solubility was similar in the various types of tissue. There were no basic differences between rabbit and human tissues. In contrast to short-chain primary alcohols (methanol, ethanol), 3-methylbutanol-1 was distributed in aqueous tissues as well as in anhydrous tissues. From this it was deduced that fatty tissue (water content approx. 10–20%) acts as a depot for longer-chain alcohols.

The influence of 3-methylbutanol-1 (concentration range 140–320 µmol/ml, approx. 12–28 mg/ml) on O_2 intake, CO_2 production, acid formation and lactate and pyruvate concentration was studied using liver sections obtained from 200–300 g male and female Wistar rats. No effect on oxygen intake was observed. CO_2 production was reduced and acid production increased. In addition, there was a significant increase in the lactate/pyruvate concentration ratio (Forsander, 1967).

7.2 Acute and subacute toxicity

5 male and 5 female mice each received intraperitoneally a dose of 700 mg 3-Methylbutanol-1 kg. The observation period was 14 days. In this period, 3 males and 3 females died (LD_{50} i.p. approx. 700 mg/kg). Toxic symptoms observed were dyspnoea, apathy, impaired balance, atonia, pareses of the rear extremities, redness of the skin, scaliness and poor general condition (BASF, 1979).

For female mice (breed H, 7–8 weeks old) the intravenous LD_{50} was 2.64 mmol/kg (approx. 233 mg/kg; Chvapil et al., 1962).

5 male and 5 female Sprague-Dawley rats were each administered 2150 and 5000 mg/kg by stomach tube. The observation period was 14 days. In this period, each of the doses of 2150 and 5000 mg/kg were lethal to one female. The LD_{50} value was thus in excess of 5000 mg/kg. Toxic symptoms observed were dyspnoea, apathy, staggering, atonia, pareses of the rear extremities and poor general condition (BASF, 1979).

In the range-finding test, Smyth et al. (1969) determined the acute oral LD_{50} for rats as 7.07 ml/kg (4.82–10.4 ml/kg).

Purchase (1969) gave the oral LD_{50} for female rats as 4.0 g/kg (2.45–6.17 g/kg) and for males as 1.30 g/kg (0.67–2.41 g/kg; no details of strain). The observation period was 10 days. Under dissection, there were indications of degenerative changes in the liver and the kidneys.

The dose of 3-methylbutanol-1 which led to muscular incoordination (graded 3 on a 7-point scale of increasing neurological symptoms) when administered intraperitoneally to male Sprague-Dawley rats (250–350 g) was 23 mmol/kg (corresponding to around 2 g/kg; Shoemaker, 1981).

12 rats (no details of sex or strain) inhaled an enriched atmosphere of 3-methylbutanol-1 at 20° C over 7 hours. None of the exposed animals died. Toxic symptoms were discribed as panting and loss of the pain reflex (BASF, 1979).

Rats tolerated inhalation of an atmosphere saturated with 3-methylbutanol-1 at room temperature (range-finding test) over 8 hours (Smyth et al., 1969).

Lendle (1928) determined the narcotic dose for 6 rats (no details of sex or strain) following intraperitoneal administration of 3-methylbutanol-1 (administered as a 2% solution in physiological NaCl solution) as 0.5 ml/rat, and the lethal dose as 1.0 ml/rat.

Haggard et al. (1945) gave the lethal dose on intraperitoneal administration (the dose which produced cessation of breathing) as 610 mg/kg (no further details).

The LD_{50} of 3-methylbutanol-1 for rabbits (both sexes, 1.5–2.5 kg) following oral administration was determined as 39 mmol/kg (corresponding to 3.4 g/kg). The observation period was 24 hours. In addition, the narcotic dose ND_{50} (defined as the dose which resulted in stupor and loss of voluntary movement in half of the rabbits) was determined. The ND_{50} was 8 mmol/kg (corresponding to 0.7g/kg). In this experiment, high doses of the test substance (no details of the dose) caused loss of corneal reflex, nystagmus, dyspnoea and bradycardia (Munch, 1972).

The test substance was administered to rabbits (1–2.5 kg, no details of strain, sex or number per group) orally by stomach tube. The minimum narcotic dose (the dose which produced mild narcosis in half the animals, as shown by stupor, immobility, lying on the side or the stomach, short-term resumption of movement and standing up on manual compression or stimulation) was 0.87 ml/kg, and the lethal dose was 4.25 ml/kg (Munch and Schwarzte, 1925).

Smyth et al. (1969) gave the dermal LD_{50} for rabbits as 3.97 ml/kg (2.93–5.37 ml/kg; range-finding test, 24 hours exposure, occlusive).

Following i.v. administration, to 3 rabbits (2.5–3.7 kg, no other details), the minimal anaesthetic dose of 3-methylbutanol-1, i.e. the dose leading to loss of corneal reflex, was given as 0.85 g/animal, and the minimal lethal dose as 1.57 g/animal (Lehman and Newman, 1937).

The lethal intravenous dose of 3-methylbutanol-1 for cats was given as 0.26 ml/kg (injected as 1% solution in physiological NaCl solution, 2 ml/minute, tested on 1 cat; Macht, 1920).

The cardiovascular effects of 3-methylbutanol-1 administered intravenously by continuous infusion (20 mg/kg/minute) to 6 dogs (15–30 kg) which had been anaesthetized by sodium pentobarbital (30 mg/kg i.v.), were an increasing reduction in pulse rate per minute, a reduced systolic arterial pressure, and a reduction in the contraction rate of the cardiac muscle, even when the 3-methylbutanol-1 blood concentration was less than 0.2 mg/ml. All animals died following 1 hour's infusion (Nakano and Kessinger, 1972).

7.3 Skin and mucous membrane irritation

Following testing in accordance with Federal Register 28 No. 187 (1973) on the intact and scarified skin of rabbits (occlusive, 24 hours), 3-methylbutanol-1 was rated as moderately irritant (BASF, 1979).

On rabbits' eyes, a distinct reddening and swelling of the mucous membranes and a mist-like corneal opacity were observed, with both changes receding slowly (BASF, 1979).

In a range-finding test, only slight reddening of the skin occurred, but severe symptoms of irritation to the mucous membranes of the eye were described (Smyth et al., 1969).

7.4 Sensitization

No information available.

7.5 Subchronic-chronic toxicity

Groups of 15 male rats (80–115 g) and 15 female rats (80–105 g, Ash/CSE strain) were given daily oral doses by stomach tube (7 days a week) of 3-methylbutanol-1 at 0 (control), 150, 500 or 1000 mg/kg for 17 weeks. In addition, groups of 5 rats of each sex were given 0, 500 and 1000 mg/kg of 3-methylbutanol-1 for 3 or 6 weeks. The alcohol (at least 98% pure, 20 ppm of heavy metals) was administered dissolved in corn oil at a volume of 5 ml/kg/day. The animals were weighed regularly once a week up to day 110 of the study, with recording of food and water consumption prior to the day of weighing. At the maximum dose (1000 mg/kg), there was a delayed increase in weight as a result of reduced food absorption. In comparison with the controls, no effects were evident on analysis of the blood (Hb, erythrocytes, reticulocytes, leucocytes and differential blood count) or the urine (urea, glucose, total albumin, albumin) or in biochemical tests (GOT, GPT, LDH). Histopathological examination of the organs tested (brain, heart, liver, spleen, kidneys, stomach, gastro-intestinal tract, adrenal glands, gonads, pituitary gland, thyroid gland) also revealed no changes attributable to 3-methylbutanol-1 (Carpanini et al., 1973).

7.6 Genotoxicity

No information available.

Table 1. Test methods and results of investigations on the carcinogenic effect of 3-methylbutanol-1

Group	Route of application	Animals/group	Individual dose (ml/kg)	Mean total dose (ml)	Mean survival time (days)	Animals with tumours					
						Total		Maligne		Benigne	
						n	(%)	n	(%)	n	(%)
Control[a]	oral	25	1.0[b]	–	643	3	12	–	–	3[d]	12
Control[a]	s.c.	25	1.0[b]	–	643	2	8	–	–	2[d]	8
3-Methyl-butanol-1	oral	15	0.1[b]	27.0	527	7	47	4[f]	27	3[e]	20
3-Methyl-butanol-1	s.c.	24	0.04[c]	3.8	592	15	62	10[g]	42	5[e]	21

[a] 0.9% NaCl solution
[b] Twice a week
[c] Once a week
[d] Papillary tumours and papillomatoses of the forestomach, fribroadenomas of the mammary gland
[e] Not defined for this group
[f] 1 myeloid leukaemia, 2 liver cell carcinomas, 1 forestomach carcinoma
[g] 4 liver sarcomas, 2 liver carcinomas, 1 spleen sarcoma, 1 carcinoma of the forestomach, 2 myeloid leukaemia

7.7 Carcinogenicity

In a carcinogenicity study not conforming to present-day requirements, 3-methylbutanol-1 (degree of purity doubly-distilled, no other details) was administered to male and female Wistar rats (colony-bred) orally by stomach tube (twice a week) or subcutaneously (once a week). It was not stated whether the alcohol was administered undiluted or in a preparation. In each case, 30 Wistar rats (30 oral, 30 subcutaneous, no details of distribution by sex) served as controls and were given physiological saline solution. All animals were observed until they died naturally, then dissected and prepared histologically. The blood picture (leucocytes, differential blood count) was also analyzed. Specific details of the results of these investigations were not given, or only in general terms. The following were listed as chronic-toxic lesions in the treated animals: hyperplasia of the blood-forming parenchyma of the bone marrow, leukaemic infiltration in the liver and kidneys (all 3 haemopoietic cell systems were affected), toxic liver damage (congestion, cell necrosis, fibrosis, cirrhosis), localized scar formation in the heart muscle, and occasional interstitial pancreatitis and fibrosis (Gibel et al., 1975).

A summary of the tumours found is given in table 1.

The absolute number of tumours was higher than for the controls both for oral and s.c. administration; the tumours were distributed among various organ systems.

The work referred has to be viewed critically because of:
- Insufficient data on the purity of the alcohol used.
- The small number of animals used limits interpretation of the results. No data on tumours for historical controls.
- No range-finding studies.
- Since only one dose was administered, it is not possible to establish a dose-response relationship.
- The dose used in each case produced chronic-toxic lesions, so it may be assumed that the dose chosen was above the usual doses for carcinogenicity tests.
- The documentation of the global findings does not permit evaluation.

Consequently, this study cannot be evaluated.

7.8 Reproductive toxicity

The McLaughlin working group (1964) tested the effect of 3-methylbutanol-1 on chick embryos. Various doses of the substance were injected into the yolk sacs of fertilized chick embryos, which were then incubated. At a dose of 8 mg/egg chick hatched from 85% of the eggs, at 16 mg/egg 40% hatched, and at 40 mg/egg no chicks hatched. No deformities were observed in the hatched chicks.

7.9 Effects on the immune system
No information available.

7.10 Other effects
No information available.

8. Experience in humans

3-Methylbutanol-1 occurs in varying amounts in alcoholic drinks. In a drinking test (no details of method), test subjects (number not specified) took 40% by volume of ethanol and 3.75 mg/l of 3-methylbutanol-1 in orange juice (total quantity not stated). Isovaleraldehyde and isovaleric acid were detected in the blood as metabolites of 3-methylbutanol-1. No unchanged 3-methylbutanol-1 was detectable in the blood (observation over 9 hours from the end of drinking). The metabolite concentration was at its maximum at the end of drinking (period of drinking not stated), and then declined continuously over the period of observation (up to 9 hours later; Rüdell et al., 1983).

In a drinking test by the same investigators, in which 10 test subjects were given 40% by volume of ethanol in orange juice alone and with the addition of 3-methylbutanol-1 (1 g/l; no details of the quantity drunk), the following tests were undertaken before commencement of drinking, and 1 and 9 hours after the end of drinking: testing reaction times, two-handed coordination tests, d_2 tests (attention loading tests) and subjective judgement. In the acute alcoholization phase, there was a good correlation between the expected determination in performance and the blood ethanol concentration. In the post-alcoholic phase, a considerable increase in the number of errors and subjective symptoms of hangover were recordet in tests with higher alcohols, including 3-methylbutanol-1 (Rüdell et al., 1981).

Nelson et al. (1943) determined the concentration of 3-methylbutanol-1 found to be irritating by the majority of subjects (test subjects of both sexes, 10 subjects in total), in an inhalation chamber with exposure periods of 3–5 minutes; this were 150 ppm for the eyes and the nose and 100 ppm for the throat.

The odour threshold for 3-methylbutanol-1 in a test fluid (94.4% alcohol distilled from grain) was found to be 7 ppm (Salo, 1970).

The odour threshold of 3-methylbutanol-1 in the atmosphere has been given as 0.042 ppm (Amoore and Hautala, 1983).

References

Amoore, J.E., Hautala, E.
Odor as an aid to chemical safety: Odor thresholds compared with threshold limit values and volatilities for 214 industrial chemicals in air and water dilution
J. Appl. Toxicol., 3, 272–290 (1983)

BASF AG
Investigation of the acute toxicity of 3-methylbutanol-1
Report dated 26.6.1979

BASF AG
Investigation of the acute skin irritation of 3-methylbutanol-1
Report dated 26.6.1979

Carpanini, F.M.B., Gaunt, I.F., Kiss, I.S., Grasso, P.
Short-term toxicity of isoamyl-alcohols in rats
Fd. Cosmet. Toxicol., 11, 713–724 (1973)

Chvapil, M., Zahradnik, R., Cunchalova, B.
Influence of alcohols and potassium salts of xanthogenic acid on various biological objects
Arch. Int. Pharmacodyn., 135, 330–343 (1962)

Deutsche Forschungsgemeinschaft
Maximum Concentrations at the Workplace and Biological Tolerance Values for Working Materials 1986
Report No. XXIII for the investigation of Health Hazards of Chemical Compounds in the Work Area
Verlag Chemie, Weinheim, 1987

Forsander, O.A.
Influence of some aliphatic alcohols on the metabolism of rat liver slices
Biochem. J., 105, 93–97 (1967)

Gaillard, D., Derache, R.
Vitesse de la métabolisation de différents alcools chez le rat
C. R. Séance Soc. Biol., 158, 1605–1608 (1964)

Gaillard, D., Derache, R.
Métabolisation de différents alcools, présents dans les boissons alcooliques chez le rat
Tr. Soc. Pharmacie Montpellier, 25, 51–62 (1965)

Gaillard, D., Derache, R.
Action de quelques alcools alipathiques sur la mobilisation de différentes fractions lipidiques chez le rat
Fd. Cosmet. Toxicol., 4, 515–520 (1966)

Gibel, W., Gobs, K.H., Wildner, G.P., Schramm, T.
Experimentelle Untersuchungen zur kanzerogenen Wirkung höherer Alkohole am Beispiel von 3-Methyl-1-butanol, 1-Propanol und 2-Methyl-1-propanol
Z. Exp. Chir. 7, 235–239 (1974)

Gibel, W., Gobs, K.H., Wildner, G.P.
Experimentelle Untersuchung zur kanzerogenen Wirkung von Lösungsmitteln am Beispiel von Propanol-1, 2-Methylpropanol-1 und 3-Methylbutanol-1
Arch. Geschwulstforsch., 45, 19–24 (1975)

Greenberg, L.A.
The appearance of some congeners of alcoholic beverages and their metabolites in blood
Quart. J. Stud. Alcohol, Suppl. 5, 20–25 (1970)

Haggard, H.W., Miller, D.P., Greenberg, L.A.
The amyl alcohols and their ketones: their metabolic fates and comparative toxicities
J. Ind. Hyg. Toxicol., 27, 1–14 (1945)

Hedlund, S.G., Kiessling, K.-H.
The physiological mechanism involved in hangover
Acta Pharmacol. Toxicol., 27, 381–396 (1969)

Kamil, I.A., Smith, J.N., Williams, R.T.
Studies in detoxication
Biochem., 53, 129–136 (1953)

Kühnholz, B., Wehner, H.D., Bonte, W.
In-vitro-Untersuchungen zur Löslichkeit aliphatischer Alkohole in Körpergeweben
Blutalkohol, 21, 308–318 (1984)

Lehmann, A.J., Newman, H.W.
Comparative intravenous toxicity of some monohydric saturated alcohols
J. Pharmacol. Exp. Ther., 61, 103–106 (1937)

Lendle, L.
Beitrag zur allgemeinen Pharmakologie der Narkose
Arch. Exp. Pathol. Pharmacol., 129, 214–245 (1928)

Macht, D.I.
A toxicological study of some alcohols with especial reference to isomers
J. Pharmacol. Exp. Ther., 16, 1–10 (1920)

McLaughlin, J., Marliac, J.P., Verrett, M.J., Mutchler, M.K., Fitzhugh, O.G.
Toxicity of fourteen volatile chemicals as measured by the chick embryo method
Ind. Hyg. J., 25, 282–284 (1964)

Munch, J.C., Schwarzte, E.W.
Narcotic and toxic potency of aliphatic alcohols upon rabbits
J. Lab. Clin. Med., 10, 985–996 (1925)

Munch, J.C.
Aliphatic alcohols and alkyl esters: narcotic and lethal potencies to tadpoles and to rabbits
Ind. Med., 41, 31–33 (1972)

Nakano, J., Kessinger, J.M.
Cardiovascular effects of ethanol, its congeners and synthetic bourbon in dogs
Eur. J. Pharmacol., 17, 195–201 (1972)

Nelson, K.W., Ege, J.E., Ross, M., Woodman, L.E., Silverman, L.
Sensory response to certain industrial solvent vapor
J. Ind. Hyg. Toxicol., 25, 282–285 (1943)

Purchase, I.F.H.
Studies in Kaffircorn malting and brewing
S. A. Med. J., 54, 795–798 (1969)

Rüdell, E., Bonte, W., Spring, R., Frauenrath, C., Küssner, H., Sellin, I.
Pharmakologische Wirkungen geringer Dosen höherer alipathischer Alkohole
Blutalkohol, 18, 315–325 (1981)

Rüdell, E., Bonte, W., Sprung, R., Kühnholz, B.
Zur Pharmakokinetik der höheren alipathischen Alkohole
Beitr. Gerichtl. Med., 41, 211–218 (1983)

Salo, P.
Determining the odor thresholds for some compounds in alcoholic beverages
J. Food Sci., 35, 95–99 (1970)

Shoemaker, W.J.
The neurotoxicity of alcohols
Neurobehav. Toxicol. Teratol., 3, 431–436 (1981)

Smyth, H.F., Carpenter, C.P., Weil, C.S., Pozzani, U.C., Strigel, J.A., Nycum, J.S.
Range-finding toxicity data, List VII
Arch. Ind. Hyg. Ass. J., 30, 470–476 (1969)

Ullmanns Encyclopaedia of Industrial Chemistry
4. Edition, Vol. 7, p 531
Verlag Chemie, Weinheim, 1974

Weast, R.C. (ed.)
CRC Handbook of Chemistry and Physics, 62. edition, C-192
CRC Press, Boca Raton, USA, 1981/83

Windholz, M. (ed.)
The Merck Index, 10. edition, p 747
Merck & Co., Inc., Rahway, USA, 1983

Tributyl phosphate

1. Summary and assessment

Tributyl phosphate (TBP) is rapidly metabolized by the microsomal enzyme complex in the liver to dibutyl(hydroxybutyl)phosphate, which is then metabolized to butyl-di(hydroxybutyl)phosphate and dibutyl phosphate. Irrespective of the route of administration, elimination occurs principally in the urine (50–70%) but also in the breath (7–10%) and the faeces (4–6%).

In terms of lethal effects, TBP is of low acute toxicity (oral LD_{50} in the rat 1400–3000 mg/kg). The marked difference in effect of a single intravenous injection, compared with oral and dermal administration, indicates the rapid metabolism of this substance, which occurs in the gastrointestinal tract and in the skin. The observed symptoms show TBP to have an effect on the central nervous system (narcosis) and the respiratory system (dyspnoea, apnoea, lung oedema) as well as a neurotoxic effect (muscular paralysis).

This substance is strongly irritating to the skin and mucosa, including the lining of the respiratory tract, both in animals and in man.

Repeated administration of TBP results in definite effects on the liver and kidneys, with increases in their weight and signs of functional disturbance (raised transaminase and cholesterol values and a raised blood-urea concentration). Furthermore, pathological changes in the bladder epithelium, ranging from diffuse hyperplasia to focal nodular hyperplasia, may develop. In other experiments, liver-cell necrosis and dystrophic changes in the renal tubules are found.

TBP is not mutagenic in vitro or in vivo.

The substance does not affect the fertility of male animals. A weak teratogenic potential has been noted on direct contact with embryos (tests on fertilized hens' eggs).

TBP has demonstrated neurotoxic properties. In rats this involves neurochemical dysfunction, reduction of nerve-conduction speed, slight inhibition of cholinesterase activity and morphological changes in nervous tissue (changes in Schwann cells, hyperaemia of the brain and disintegration of cortical cells); however, there is no detailed information. No corresponding effects have been observed in chickens.

In humans, exposure to TBP by inhalation leads to nausea and headache, in addition to the irritative effects already noted. A low-grade inhibition of cholinesterase in plasma and erythrocytes has been observed in in vitro experiments on human blood.

The threshold limit value is 2.5 mg/m^3 of air.

2. Name of substance

2.1 Usual name   Tributyl phosphate
2.2 IUPAC name   Phosphoric acid tri-n-butyl-ester
2.3 CAS-No.   126-73-8

3. Synonyms

Common and trade names

Butyl phosphate
Tri-n-butyl phosphate
Celluphos 4
Disflamol 1 TB
TBP
Phosflex 4
Skydrol LD-4

4. Structural and molecular formulae

4.1 Structural formula

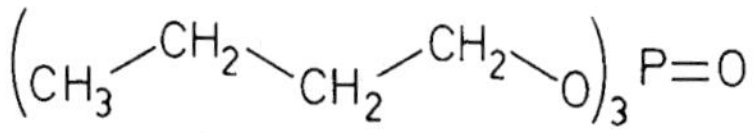

$$\left(CH_3 - CH_2 - CH_2 - CH_2 - O\right)_3 P=O$$

4.2 Molecular formula $C_{12}H_{27}O_4P$

5. Physical and chemical properties

5.1 Molecular mass   266.36 g/mol
5.2 Melting point   <–80° C
5.3 Boiling point   289° C (decomposition)
5.4 Density   0.982 g/cm^3 (at 20° C)

5.5 Vapour pressure	0.09 hPa (at 25° C)
	9.7 hPa (at 150° C)
	169.3 hPa (at 177° C)
5.6 Solubility in water	1 ml in 165 ml water
5.7 Solubility in organic solvents	Readily soluble in alcohol, ether
5.8 Solubility in fat	No information available
5.9 pH-value	–
5.10 Conversion factors	1 ppm $\hat{=}$ 11.05 mg/m^3 air
	1 mg/m^3 air $\hat{=}$ 0.09 ppm

(CHIP, 1985; Sax, 1984)

6. Uses

Plasticiser for plastics, paint component, foaming agent, flame-retarding agent in hydraulic fluids, solvent and extraction agent for heavy metal ions (e.g. PUREX process) (CHIP, 1985; Ullmann, 1979)

7. Results of experiments

7.1 Toxicokinetics and metabolism

The rate of metabolism of TBP and the metabolites produced were determined in in vitro tests on rat liver homogenate. It was found that rat liver microsome enzymes rapidly metabolise TBP in the presence of NADPH (within 30 minutes), with only slight metabolic breakdown (11%) in the absence of added NADPH. Dibutyl-(hydroxybutyl)phosphate was obtained as a metabolite in the first stage of the test. The extended incubation time in the second stage of the test yielded a further 2 metabolites, butyl-di-(hydroxybutyl)phosphate and dibutyl hydrogen phosphate, which are produced from the primary metabolite dibutyl(hydroxybutyl)phosphate (Sasaki et al., 1984).

Male Wistar rats (weighing 180–210 g) received a single oral or intraperitoneal dose of 14 mg ^{14}C-labelled TBP/kg body weight (specific activity 0.179 mCi/mmol) dissolved in 0.1 ml of oil. Urine and faeces were collected separately and after appropriate processing the radioactivity was determined quantitatively by liquid scintillation. This showed that within 24 hours of oral administration, 50% of the radioactivity is eliminated in the urine, 10% in the exhaled air and 6% in the faeces, the total elimination after 5 days being 82%. Following intraperitoneal injection, 70% of the radioactivity was eliminated in the urine, 7% by exhalation and 4% in the faeces within 24 hours, 90% being eliminated after 5 days. A total of 11 metabolites were found in the 24-h urine folowing a single intraperitoneal injection of 250 mg TBP. The main metabolites were dibutyl hydrogen phosphate, butyl dihydrogen phosphate and butyl-bis-(3-hydroxybutyl)-phosphate. Dibutyl-(3-hydroxybutyl)phosphate and dibutyl-(3-oxobutyl)phosphate occurred as intermediates (Suzuki et al., 1984).

7.2 Acute and subacute toxicity

With respect to acute toxicity TBP has proved to be slightly toxic. The oral LD_{50} on rats lay between 1400 and 3200 mg/kg (Johannsen et al., 1977; Eastman Kodak, 1958; Mitomo et al., 1980, Bayer, 1986a), on mice from 400 to 1240 (Eastman Kodak, 1958; Mitomo et al., 1980), and on hen of 1800 mg/kg body weight (Johannsen et al., 1977). The dermal LD_{50} in rabbits was more than 3100 mg/kg body weight (Johannsen et al., 1977), the subcutaneous LD_{50} was approximately 3000 mg/kg body weight (mice). The intraperitoneal LD_{50} lay between 500 and 1600 mg/kg body weight (rats, mice). The LC_{50} following inhalation was determined in cats as 2500 mg/m^3 air (4 to 5 hours exposure), and more than 1359 mg/m^3 air in rats (6 hours exposure). An intravenous injection of about 100 mg/kg body weight (rats) was lethal. Clinical symptoms such as narcosis, cyanosis, twitching, dyspnoea, and respiratory arrest (with pulmonary oedema) were observed.

The oral administration of 1 ml TBP/kg body weight over 4 days was fatal to 2 guinea pigs (group size 3 animals, no further information; Patty, 1963).

Doses of 0.14 and 0.42 ml TBP/kg body weight were administered daily to each of 10 male and female Sprague-Dawley rats (weighing roughly 225 and 177 g) by stomach tube for 14 days. Control animals received 0.42 ml of water/kg body weight. 24 hours after the last administration, blood was withdrawn from the vena cava caudalis of 6 animals (under ether narcosis) in each group for the determination of red and white blood cell counts and clinical-chemical tests including acetylcholinesterase (corresponding to the standard OECD programme). The following organs were removed, weighed and fixed in 10% phosphate-buffered formalin (pH

7.0) for gross and histopathological examination: brain, heart, kidneys, liver, lungs, spleen, ovaries and testes. Sections prepared from the organs embedded in paraffin were stained with haematoxylin-eosin and examined under a light microscope. Treatment produced no overt toxic symptoms. The body weight gain of the treated animals corresponded to that of the controls. With the exception of a significantly reduced haemoglobin content in the females of the higher dose group, the red and white blood cell counts for the treated animals were comparable with those of the controls. Clinical-chemical investigations showed increased potassium content in females of both dose groups, and increased amylase and triglyceride levels in the higher dose group. In males, amylase and bilirubin levels were increased in the higher dose group. Liver weight was increased in the higher dose group in both sexes ($p < 0.001$). The females in this group also had a lower spleen weight ($p < 0.05$). Macroscopic examination of the organs showed a slightly enlarged liver and a slightly smaller spleen in the higher dose group. Degenerative changes in the sperm duct (aspermia, giant cells, cells with pycnotic or karyorrhectic nuclei) were found in one (male) animal of the 4 animals in the higher dose group that were subjected to histopathological examination. There were no other histopathological findings (Laham et al., 1984).

Male JCL-Wistar rats (weighing 65–75 g, age 4 weeks) received 0 or 0.5% TBP in the food (pellets) for 9 weeks. At the end of the treatment period, blood (citrate buffered) was taken from the vena cava caudalis of the animals under ether narcosis and the liver, kidneys, spleen and testes were weighed. A red blood cell count was made including prothrombin and partial thromboplastin times, and protein, urea, cholesterol and enzyme activities (ALT, AST and ALP) were determined in the serum. Sodium and potassium contents were determined by flame spectrometry. The liver, kidneys and spleen were subjected to histopathological examination. The animals treated with TBP showed a lower body weight (−11%) than the control animals, and absolute and relative liver weights were significantly increased ($p < 0.05$). A significantly increased urea content was measured in the blood ($p < 0.05$). The other parameters and histopathological investigations showed no treatment-related effects (Oishi et al., 1982).

In an investigation extending over 10 weeks rats which had received 0.5 or 1% TBP in the food showed an increased prothrombin time and partial thromboplastin time which were not confirmed in the study mentioned before. The other changes were comparable (Oishi et al., 1980).

Seven consequtive daily intubation of TBP for rats in doses of 0.14 and 0.20 (0.98 and 1.40 g/kg in total) resulted in marked increments of relative weights of liver and kidneys in acompany with an increase of BUN value and tubular degeneration. The daily intubation of 0.13 and 0.46 g/kg of TBP for one month caused in rats a marked depression of body weight gain and lethal cases by 20 and 40%, respectively. Tubular damage was also a major change (Mitomo et al., 1980).

7.3 Skin and mucous membrane effects

Severe skin and mucosal irritation has been observed in rabbits following contact with liquid TBP (nor further information; ACGIH, 1980).

A single dermal application of 0.5 g TBP to the intact or abraded skin of 6 rabbits (no further information) produced severe irritation, inducing erythema and oedema in all the animals (FMC Corporation, 1985a).

A test on the irritating and corrosive potential of TBP according to "OECD Guideline for Testing of Chemicals No. 404 and 405" showed that TBP was slightly irritating to rabbit skin (4-hour exposure) and to rabbit eyes (Bayer, 1986b).

The instillation of 0.1 g TBP in the conjunctival sac of rabbits caused mild irritation 24, 48 and 72 hours as well as 7 days following application (no further information; FMC Corporation, 1985a).

7.4 Sensitization

After dermal treatment of 14 guinea-pigs in a "droplet-test" (procedure and evaluation not described) 6 animals showed signs of skin-sensitizing effects (no further details; Eastman Kodak, 1958).

7.5 Subchronic-chronic toxicity

12 male and 12 female Sprague-Dawley rats (average weight 206 and 294 g) received 0.20 and 0.30 g/kg body weight daily (0.35 g/kg from the 7th week in the latter case) administered by stomach tube for 18 weeks. Control animals received 0.35 g water/kg body weight. During the treatment period the animals were observed daily for clinical symptoms. Body weight gain was measured weekly. At the end of the treatment period, blood was obtained from the vena cava caudalis of 6 animals (under ether narcosis) for each dose, for haematological and clinical-chemical investigations (further to the

OECD standard). 6 animals for each dose were examined macroscopically following dissection and the following organs were removed for histopathological examination: adrenals, brain, heart, kidneys, liver, lungs, gastrointestinal tract, spleen, sex organs, thyroids and bladder. The heart, kidneys, liver, lungs, spleen and sex organs were weighed. All the organs were fixed in formalin (pH 7) and embedded in paraffin. The sections were stained with haematoxylin-eosin and examined under a light microscope. No clinical symptoms were observed during the study. In comparison with controls, a reduced body weight gain (of some 10%) occurred in the treated males from the third week and persisted until the end of the test. Red and white blood cell counts showed no significant changes. The clinical-chemical findings revealed a very slight inhibition of acetyl-cholinesterase activity in the females in the higher dose group. The absolute and relative liver and spleen weights of the females and the relative kidney weights of the males were increased in the high dose group. Histopathologically, all the treated animals showed diffuse hyperplasia of the bladder epithelium, associated with subepithelial hyperplasia. Focal nodular hyperplasia of the epithelium and slight mononuclear infiltration were observed predominantly in the males. No other significant changes were found (Laham et al., 1985).

JCL-Wistar rats and DDY mice of both sexes received 0.05, 0.2 and 1% TBP in the food for 3 months. A dose-related reduction in body weight gain and lower uterus weights were observed in both species. Liver, kidney and testicle weights increased. An increased blood urea content was found in the highest dose group (no further information; Mitomo et al., 1980).

15 male and 15 female Sprague-Dawley rats were administered TBP in the food for 13 weeks in the following concentrations: 0, 8, 40, 200, 1000 and 5000 ppm daily. Reduced body weight gain and food consumption occurred above 1000 ppm. No deaths occurred. Of the clinical-chemical parameters, an increase in prothrombin time and partial thromboplastin time, increased ALT and AST activity and increased cholesterol and calcium levels were observed above 1000 ppm. Hyperplasia of the bladder epithelium was observed on histopathological examination (Cascieri et al., 1985).

Liver necrosis was observed in further investigations in rabbits and rats which received 0.2–5 mg TBP/kg daily (no further information; Zyabbarova and Teplyakova, 1968).

Liver necrosis, associated with increased liver weight, was observed in rats in another test, in addition to increased kidney weights and tubular dystrophy (no further information; Pupyshava and Peresedov, 1970).

In a test involving repeated administration to rats, Kalinina (1970) observed increased bromosulphthalein retention (no further information).

7.6 Genotoxicity

TBP was tested for mutagenic effects in a Salmonella/microsome test with and without S9-mix (metabolising system) at doses of up to 12,500 µg/plate on four Salmonella typhimurium LT2 mutants. Histidine-auxotrophic strains TA 1535, TA 100, TA 1537 and TA 98 were used. Doses of up to 120 µg/plate produced no bacteriotoxic effects. Bacterial counts remained unchanged. In the higher dose range there was marked strain-specific bacterial toxicity so that only the range up to 500 µg/plate could be evaluated. The investigation showed no indications of any mutagenic effect by the substance (Bayer, 1985).

The testing of 5 to 10 µl of TBP/plate on Salmonella typhimurium strains hisC337, hisG46, TA 1530 and TA 1534 without a metabolising system (corresponding to the standard protocol) yielded no mutagenic effect after 48 and 72 hours' incubation at 37° C (Hanna and Dyer, 1975).

The testing of TBP at doses of 97-97,000 µg/plate with and without a metabolising system (S9-mix) on Salmonella typhimurium strains TA 98, TA 100, TA 1537 and TA 1538 also showed no mutagenic activity (no further data; FMC Corporation, 1985b).

A dose of 9.8 µg TBP/plate did not lead to an increase of vevertants on the Salmonella strain TA 102, cohen S9-MIX was added (Pancorbo et al., 1987). In another Salmonella/microsome test, where Salmonella strains TA 1535 and 1538 were incubated with 500 and 1000 µg TBP/plate an 3- to 7 fold increase was found, when S9-Mix was added (Gafiea and Chudin, 1986).

Tests on Escherichia coli strains WP2, WP2uvrA, CM561, CM571, CM611, WP67 and WP12 (standard protocol) likewise showed no mutagenic effect after 48 and 72 hours' incubation at 37° C (Hanna and Dyer, 1975).

TBP was investigated for an in vivo mutagenic effect on male Drosophila flies. The animals received TBP in the food at concentrations which had no insecticidal effect. They were then mated with untreated females and up to 10 male progeny per generation (F1 to F4) were investigated for infertility and recessive-linked lethality. The males showed no such effects on oral uptake of TBP, and thus showed no evidence of a mutagenic effect (Hanna and Dyer, 1975).

7.7 Carcinogenicity

No information available.

7.8 Reproductive toxicity

As described in 7.6, the Drosophila test showed no indications of infertility in male animals or of teratogenic effects in male offspring (Hanna and Dyer, 1975).

As described in 7.2, degeneration of the sperm duct was observed in one male rat (4 rats investigated) following 14 days of treatment with 0.42 ml TBP/kg body weight (Laham et al., 1984).

5 mg TBP was applied to the yolk sac of 10 fertilized Leghorn chickens and 18–21 day old embryos were investigated for malformations. Weight, body length and leg length were determined in the embryos, together with beak and leg anatomy, feathering and rate of hatching. The treatment had a slight teratogenic effect in the form of reduced body weight gain, reduced body and leg length and reduced hatching rate (−20%; Roger et al., 1969).

7.9 Effects on the immune system

No information available.

7.10 Other effects

Male Swiss-Albino mice (weight 54–77 g) received a single intraperitoneal injection of 850–1000 mg TBP/kg (formulated in oil). The treatment gave rise to muscular paralysis and associated breathing difficulties. The average knockdown time (time from injection to immobilisation of the animals) was 11.9 (±0.8) minutes. Tests with different nicotinic acid derivatives and pyridine compounds for an antagonistic action suggested that the effect of TBP involves blocking the transfer of impulses at neuromuscular connections (Chambers and Casida, 1967).

Rats which received 3000 mg TBP/kg orally or 800–1600 mg TBP/kg by intraperitoneal injection did not show this effect (no further information; Patty, 1963).

Hens (white Leghorns) received 1840 mg TBP/kg by stomach tube on 2 days so that the cumulative dose was greater than the LD_{50} dose. The animals were observed after 42 days, with macroscopic examination following sectioning and histopathological examination of the brain, ischiac nerve and spinal cord after appropriate staining. No neurological symptoms were observed. Histopathological examination of the removed tissue showed no indications of a neurotoxic effect of TBP on hens (Johannsen et al., 1977).

10 male and 10 female Sprague-Dawley rats (weighing 225 and 177 g) received 0.28 and 0.42 ml TBP/kg daily by stomach tube for 14 days. The control animals received 0.42 ml water/kg. The rats were observed daily for indications of clinical symptoms and body weight was determined weekly. Physiological measurements were made of conduction speed and absolute and relative refraction periods in the caudal nerve of 4 animals/dose. Two weeks after the last treatment, the two ischiac nerves of the control animals and the high-dose animals (0.42 ml) were removed after perfusion and, following appropriate fixing and embedding of transverse and longitudinal sections, were examined under the light microscope and the electron microscope. The 14-day treatment was tolerated without clinical symptoms. After 7 days all the treated animals showed reduced body weight gain compared to control-animals. The physiological measurements showed a significant ($p=0.048$) reduction in the rate of conduction in the high-dose males. The absolute refraction period was significantly ($0.001 < p < 0.01$) increased in all the treated males and females receiving the low dose, and the relative refraction period was increased in both sexes on the low dose. Investigation by the light microscope showed no changes in the ischiac nerve, whilst examination through the electron microscope revealed regression of the Schwann cell processes in non-myelinised nerve fibres. The authors suggested that this finding represented an initial treatment-related effect. As a result of this investigation, because of the lack of neurological symptoms and the unclear findings seen under the electron microscope, TBP was not considered to be harmful to axons (Laham et al., 1983).

TBP was tested on HeLa cells in the MIT-24 system (metabolic inhibition test with microscopic examination 24 hours after incubation) for any cytotoxic effects. The substance was added in a medium to microtitre plates with HeLa cells (5×10^{-4} cells per ml), sealed with liquid paraffin and incubated for 7 days at 37° C. 24 hours after the onset of incubation the cells were examined microscopically for survival. After 7 days the inhibition rate (IC_{50}) was determined using the phenol red test. With TBP, total inhibition occurred at a concentration of 4 mg/ml after 24 hours. 1.5 mg/ml was determined as the IC_{50}

after 7 days of incubation. This indicates that TBP has a marked cytotoxic potential (Ekwall et al., 1982).

8. Experience in humans

Sabine and Hayes (1952) demonstrated in an in vitro experiment, that TBP had slight inhibiting action on cholinesterase of human plasma.
Oishi (1980) demonstrated a dose-dependent inhibition of human serum cholinesterase activity in vitro (no further information).
Investigations on isolated human skin showed that TBP has a good penetration capacity (no further information; Marzulli et al., 1965).
Apparently TBP has an irritant effect on the eyes and respiratory tract in man (no further information; Patty, 1963). Other authors have similarly described an irritant effect on the skin and mucous membranes in man (no further information; Stauffer, 1984).
The exposure of workers to 15 mg TBP/m^3 air resulted in nausea and headaches (no further information; ACGIH, 1980).

References

ACGIH
Tributyl phosphate, p. 404
Documentation of TLV's (1980)

BAYER AG
Unpublished Report, No. 13805 (1985)

BAYER AG
Tri-n-butylphosphat
Untersuchung zur akuten oralen Toxizität an männlichen und weiblichen Wistar-Ratten
unveröffentlicher Bericht, Nr. 14478 (1986b)

BAYER AG
Tri-n-butylphosphat-Untersuchungen zum Reiz-/Ätzpotential an Haut und Auge
unveröffentlicher Bericht (1986a)

Cascieri, T., Ballester, E.J., Serman, L.R., McConnell, R.F., Thackara, J.W., Fletcher, M.J.
Subchronic toxicity study with tributylphosphate in rats
The Toxicologist, 5, 97 (1985)

Chambers, H.W., Casida, J.E.
Protection activity of nicotinic acid derivatives and their 1-alkyl-2- and 1-alkyl-6-pyridones against selected neurotoxic agents
Toxicol. Appl. Pharmacol., 10, 105–118 (1967)

CHIP
Chemical Hazard Information Profile (Draft)
Tri(alkyl/alkoxy)phosphates
U. S. Environmental Protection Agency (1985)

Dave, G., Lidman, U.
Range-finding acute toxicity in the rainbow trout and in the rat
Hydrometallurgy, 4, 201–216 (1978)

Ekwall, B., Nordensten, C., Albanus, L.
Toxicity of 29 plasticizers to HeLa cells in the MIT-24 system
Toxicology, 24, 199–210 (1982)

Eller, H.
Beitrag zur Toxikologie technischer Weichmachungsmittel
Dissertation, University of Würzburg (1937)

FMC Corporation
TBP – acute toxicity screening tests
Office of Toxic Substances, U.S. Environmental Protection Agency (1985a)

FMC Corporation
TBP mutagenicity screening test (Ames-test)
Office of Toxic Substances, U.S. Environmental Protection Agency (1985b)

Gafieva, Z. A., Chudin, V. A.
Evaluation of the mutagenic activity of tributylphosphate on Salmonella typhimurium
Gig. Sanit., 9, 81 (1986)

Hanna, P.J., Dyer, K.F.
Mutagenicity of organophosphorus compounds in bacteria and Drosophila
Muta. Res., 28, 405–420 (1975)

Johannsen, F.R., Wright, P.L., Gordon, D.E., Levinskas, G.J., Radue, R.W. and Graham, P.R.

Evaluation of delayed neurotoxicity and dose-response relationships of phosphate esters in the adult hen
Toxicol. Appl. Pharmacol., 41, 291–304 (1977)

Kalinina, N.I. (1970)
cited in CHIP (1985)

Laham, S., Szabo, J., Long, G.
Effects of tri-n-butylphosphate on the peripheral nervous system of the Sprague-Dawley rat
Drug Chem. Toxicol., 6, 363–377 (1983)

Laham, S., Long, G., Broxup, B.
Subacute oral toxicity of tri-n-butyl phosphate in the Sprague-Dawley rat
J. Appl. Toxicol., 4, 150–154 (1984)

Laham, S., Long, G., Broxup, B.
Induction of urinary bladder hyperplasia in Sprague-Dawley rats orally administered tri-n-butyl phosphate
Arch. Environ. Health, 40, 301–306 (1985)

Marzulli, F.N., Callahan, J.F., Brown, D.W.C.
Chemical structure and skin penetrating capacity of a short series of organic phosphates and phosphoric acid
J. Invest. Dermatol., 44, 339–344 (1965)

Mitomo, T., Ito, T., Ueno, Y., Terao, K.
Toxicological studies on tributylphosphate
J. Toxicol. Sci., 5, 270–271 (1980)

Oishi, H., Oishi, S., Hiraga, K.
Toxicity of tri-n-butyl phosphate with special reference to organ weights, serum components and cholinesterase activity in male rats
Toxicol. Lett., 6, 81–85 (1980)

Oishi, H., Oishi, S., Hiraga, K.
Toxicity of several phosphoric acid esters in rats
Toxicol. Lett., 13, 29–34 (1982)

Pancorbo, P.G., Lein, R.J., Blevins, R.D.
Mutagenic activity of surface water adjaunt to a nuclear fuel processing facility
Arch. Environ. Contam. Toxicol., 16, 531–537 (1987)

Patty, F.A.
Industrial hygiene and toxicology
2nd edition, Vol. II, p. 1853
Wiley & Sons, New York (1963)

Pupysheva, G.J., Peresedov, V.P. (1970)
cited in CHIP (1985)

Roger, J.C., Upshall, D.G., Casida, J.E.
Structure-activity and metabolism studies on organophosphate teratogens and their alleviating agents in developing hen eggs with special emphasis on Bidrin
Biochem. Pharmacol., 18, 373–392 (1969)

Sabine, J.C., Hayes, F.N.
Anticholinesterase activity of tributylphosphate
Arch. Ind. Hyg. Occup. Med., 6, 174–177 (1952)

Sasaki, K., Suzuki, T., Takeda, M., Uchiyama, M.
Metabolism of phosphoric acid triesters by rat liver homogenate
Bull. Environ. Toxicol., 33, 281–288 (1984)

Sax, N.I.
Dangerous Properties of Industrial Materials
6th edition, p. 2612
Van Nostrand Reinhold Company, New York (1984)

Stauffer Chemical Co.
Product Safety data sheet, 1984
Cited in: Eighteenth Report of the TSCA
Interagency Testing Committee To The Administrator,
Environmental Protection Agency, May 1986, p. 34

Suzuki, T., Sasaki, K., Takeda, M., Uchiyama, M.
Metabolism of tributyl phosphate in male rats
J. Agric. Food Chem., 32, 603–610 (1984)

Ullmanns Enzyklopädie der technischen Chemie
4th Edition, Vol. 18, p. 390
Verlag Chemie, Weinheim (1979)

Vandekar, M.
Anaesthetic effect produced by organophosphorus compounds
Nature, 179, 154–155 (1957)

Zyabbarova and Teplyakova (1968)
cited in CHIP (1985)

Trimethylphenyl-ammonium chloride

1. Summary and assessment

Trimethylphenyl ammonium chloride is quickly absorbed via the mucosa (eyes, gastro-intestinal and respiratory tracts) as well as through the intact skin.

It has a relatively high acute toxic action (LD_{50} in the rat orally 121 mg/kg, LD_{50} i.p. in the mouse <50mg/kg).

Is no irritative action on either the mucosa or the skin.

A dose-related but extremely weak mutagenic effect (maximum factor 2.4) has been found with the standard Ames test and also with pre-incubation, using strains TA 1538, TA 98 and TA 100 with S9 mix. Trimethylphenyl ammonium chloride is also negative in the HGPRT test on V79 cells and in the micronucleus test.

Trimethyl phenyl ammonium chloride shows a similar action to the cholinesterase-inhibitor physostigmine. In pharmacological experiments it causes strong contractions of the smooth musculature but with no miotic effect on the eyes. However, no cholinesterase inhibition could be demonstrated in the serum or red blood cells in experiments on dogs.

In view of the available results and the use of Trimethyl phenyl ammonium chloride as a methylating agent, further investigations are not considered to be urgent.

2. Name of substance

2.1	Usual name	Trimethylphenyl ammonium chloride
2.2	IUPAC name	Trimethylphenyl ammonium chloride
2.3	CAS-No.	138-24-9

3. Synonyms

Common and trade names	Phenyltrimethyl ammonium chloride

Trimethylphenyl-ammonium chloride

N,N,N-Trimethylbenzen-
ammonium chloride
Ammonyx 200
Trimethylanilinium chloride

4. Structural and molecular formulae

4.1 Structural formula

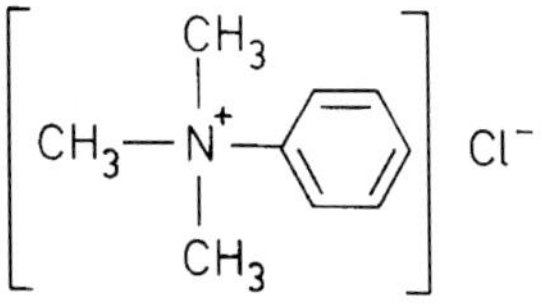

$$\left[CH_3-\overset{\overset{CH_3}{|}}{\underset{\underset{CH_3}{|}}{N^+}}-C_6H_5 \right] Cl^-$$

4.2 Molecular formula $C_9H_{14}ClN$

5. Physical and chemical properties

5.1 Molecular mass   171.69 g/mol
5.2 Melting point   234° C (Aeschliman and Reinert, 1931)
5.3 Boiling point ° C   –
5.4 Density, g/cm^3 no information available
5.5 Vapour pressure, hPa   no information available
5.6 Solubility in water   good solubility (hygroscopic) (Beilstein, 1929)
5.7 Solubility in organic solvents   good solubility in alcohol, (Beilstein, 1929) soluble in chloroform
5.8 Solubility in fat   no information available
5.9 pH-value   no information available
5.10 Conversion factors   1 ppm $\overset{\wedge}{=}$ 7.124 mg/m^3
                              1 mg/m^3 $\overset{\wedge}{=}$ 0.140 ppm

6. Uses

Methylating agent for alkaloids.

7. Results of experiments

7.1 Toxicokinetics and metabolism

According to results of studies on skin and mucous membranes (see 7.2, 7.3) as well as on the isolated rabbit intestine (see 7.10), trimethyl ammonium chloride is quickly absorbed via the mucosa (eyes, gastro-intestinal and respiratory tracts) and through the intact skin.

7.2 Acute and subacute toxicity

An LD_{50} (14-day follow-up period) of 121 (103–142) mg/kg was determined for Sprague-Dawley rats (male and female). Starting from a dose of 100 mg/kg, there was a dose-dependent occurrence of dyspnoea, apathy, staggering, trembling and clonic spasms. No pathological changes were apparent on dissection, apart from reddening of the intestinal mucosa, which was observed occasionally (BASF, 1980).

With intraperitoneal injection of 50 mg/kg to NMRI mice, 9 of the 10 animals used died within 1 hour, with convulsions (BASF, 1980).

Acute dermal toxicity was determined on rabbits; the LD_{50} (applied as 50% aqueous rub, follow-up period 14 days) was 309 (205–375) mg/kg. The animals exhibited salivation, cyanosis, dyspnoea and convulsions; locally, slight reddening of the skin was seen (BASF, 1980).

An LC_{50} (4 hours) of >0.55 mg/l was determined (exitus 0/10) for Wistar rats (5 male, 5 female/group). Trimethylphenyl ammonium chloride was administered in spray form as a 20% as well as a 60% solution in aqua bidest. The 60% solution caused death of all animals after 1 hour at a concentration of 5.6 mg/l or after 3 hours at a concentration of 2.6 mg/l. At the start of exposure there were attempts to escape, dyspnoea and convulsions; at the end of the 7-day follow-up period there was nothing unusual in the behaviour of the surviving animals. In the animals which died in the test, dissection revealed hyperaemia of the internal organs; there were no findings on dissection of the animals that were killed at the end of the test (BASF, 1987).

7.3 Skin and mucous membrane effects

An 80% aqueous preparation had a slightly irritant effect on the shaved skin of the back of white rabbits (tested according to Fed. Reg. 38, No. 187 (27.9.73), primary irritation value 2.6; BASF, 1980).

When 30 mg/animal of the powdered substance was placed in the conjunctival sac of white rabbits (White Viennese; Fed. Reg. 38, No. 187 (27.9.73)), no irritant effect on the conjunctiva was detected (primary irritation value between 2 and 3); 1 out of the 3 animals died. Instillation of 60 mg/animal into the conjunctival sac led to systemic toxic effects (dyspnoea, trembling, convulsions); all 3 rabbits died (BASF, 1980).

7.4 Sensitization
No information available.

7.5 Subchronic/chronic toxicity
No information available.

7.6 Genotoxicity
In vitro. Trimethylphenyl ammonium chloride was tested up to concentrations of 5000 µg/plate (1st experiment) and 10000 µg/plate (2nd experiment) in the Ames test with strains TA 1535, TA 1537, TA 1538, TA 98 and TA 100, with and without S9-mix. Trimethylphenyl ammonium chloride showed no bactericidal action at the concentrations tested. In both experiments for strains TA 1538 and TA 100, there was a dose-dependent increase in the number of revertants, though only by a factor of 1.4–2.4 and only in the presence of S9-mix. An increased number of revertants observed for strain TA 98 in experiment 1 could not be reproduced in the 2nd experiment. Thus, trimethylphenylammonium chloride exhibited only a very slight S9-dependent mutagenic activity. To clarify this finding, a standard Ames test with S9-mix was repeated with strains TA 1538, TA 98 and TA 100 using concentrations of up to 10000 µg/plate. A dose-dependent, but only slight mutagenic effect was found with strain TA 100 (factor 1.6) in this repeat test (Varley, 1984a).

In a preincubation test with strains TA 1538, TA 98 and TA 100 with S9-mix, only very slight dose-dependent increases in the numbers of revertants were found with a factor of 1.6 (TA 98 and TA 100) and 2.1 (TA 1538) (Varley, 1984b).

On the other hand, trimethylphenyl ammonium chloride at concentrations of 172, 573, 1146 and 1720 µg/ml did not exhibit a mutagenic effect on V79 cells in a HGPRT test with and without S9-mix (Miltenburger, 1986).

In vivo. In a micronucleus test, trimethylphenyl ammonium chloride was administered to NMRI mice as a single dose of 100 mg/kg by stomach tube. Bone marrow smears were prepared after 24, 48 and 72 hours. In none of these investigations was the incidence of micronuclei increased in comparison with the controls (Miltenburger, 1984).

7.7 Carcinogenicity
No information available.

7.8 Reproductive toxicity
No information available.

7.9 Effects on the immune system
No information available.

7.10 Other effects

Along with other structurally-related compounds, trimethylphenyl ammonium chloride was investigated for its pharmacological action in comparison with the cholinesterase-inhibitor physostigmine. In the isolated rabbit intestine, 10^{-5} molar concentration of trimethylphenyl ammonium chloride caused contraction of the smooth muscles similar in intensity to that produced by physostigmine (Aeschliman and Reinert, 1931).

A 2% trimethylphenyl ammonium chloride solution did not exhibit a miotic effect on the eye of the cat, whereas 0.1–0.5% physostigmine solution produced a definite miotic effect which persisted for several hours (Aeschliman and Reinert, 1931).

Trimethylphenyl ammonium chloride was administered orally (gelatine capsules) at doses of 1 and 10 mg/kg to groups of 2 male and 2 female beagles. An investigation of cholinesterase activity in the serum and erythrocytes was conducted 1, 3, 5, 7, 24, 48 and 96 hours later. The test substance did not produce a cholinesterase-inhibiting action, compared with a control of the same size; nor did clinical observations show any indications of increased parasympathetic tonus (Hoechst, 1986).

8. Experience in humans

No information available.

References

Aeschliman, J.A., Reinert, M.
The pharmacological action of some analogues of physostigmine
J. Pharmacol. Exp. Ther., 43, 413–444 (1931)

BASF AG
Occupational toxicological investigation of trimethylphenyl ammonium chloride
Conducted by the Department of Occupational Hygiene and Toxicology of BASF AG (1980)

BASF AG
Test Report: Acute inhalation toxicity LC_{50} (rat)
Liquid aerosol study of trimethylphenyl ammonium chloride
Department of Toxicology, dated 7.1.1987
Commissioned by the Employment Accident Insurance Fund of the Chemical Industry, Heidelberg

Beilsteins Handbuch der organischen Chemie
4th Edition, Vol. 12, No. 1601, p. 158
Springer-Verlag, Berlin (1929)

Hoechst AG
Investigation of trimethylphenyl ammonium chloride in beagle dogs to determine its inhibitory effect on cholinesterase in erythrocytes and serum
Report No. 86.1023, 1986
Commissioned by the Employment Accident Insurance Fund of the Chemical Industry, Heidelberg

Miltenburger, H.G.
Detection of gene mutations in somatic mammalian cells in culture: HGPRT-test with V79 cells
Laboratory for mutagenicity testing, Darmstadt, 26.8.1986
Commissioned by the Employment Accident Insurance Fund of the Chemical Industry, Heidelberg

Miltenburger, H.G.
Micronucleus test in bone marrow of the mouse
Laboratory for mutagenicity testing, Darmstadt, 10.8.1984
Commissioned by the Employment Accident Insurance Fund of the Chemical Industry, Heidelberg

Varley, R.
Bacterial mutagenicity tests on trimethylphenyl ammonium chloride (Report 108/8406)
Toxicol. Laboratories Limited, Ledbury, Herefordshire, England (1984a)

Commissioned by the Employment Accident Insurance Fund of the Chemical Industry, Heidelberg

Varley, R.
Further bacterial mutagenicity tests on trimethyl ammonium chloride (Report 167/8409)
Toxicol. Laboratories Limited, Ledbury, Herefordshire, England (1984b)
Commissioned by the Employment Accident Insurance Fund of the Chemical Industry, Heidelberg

Chloroformic acid ethyl ester

1. Summary and assessment

Chloroformic acid ethyl ester (CAEE) must be rated as relatively toxic (oral LD_{50} in the rat 240–400 mg/kg, LC_{50} 653 or 765 mg/m^3) and as corrosive. It quickly produces strong irritation or burning of tissue when applied to the skin. The toxic effect is, in particular, expressed in this localized tissue reaction; experiments so far reported have found no definite toxic symptoms due to absorption.

Repeated inhalation leads to irritative symptoms in the respiratory tract, breathing difficulties and swollen, haemorrhagic lungs.

In a Salmonella/microsome test with and without metabolic activation (S9-mix) CAEE produces no increase in the reversion rate even at the highest tolerable concentration of 0.4 µl/plate (with S9-mix) or 0.015 µg/plate (without S9-mix).

Long term studies with dermal and subcutaneous application as well as inhalative exposure have not shown significant hints at a tumorigenic efficiency. A dermal two stage study where CAEE has been used as initiator and phorbolmyristate acetate as promoter has revealed a marginally significant incidence of tumours.

Irritation of the eyes and respiratory tract, cyanosis and lung oedema have been observed in humans following accidental exposure to CAEE.

2. Name of substance

2.1	Usual name	Chloroformic acid ethyl ester
2.2	IUPAC name	Chloroformic acid ethyl ester
2.3	CAS-No.	541-41-3

3. Synonyms

Common and trade names Ethylchloroformiate
Ethylchlorocarbonate
Ethylchloromethanate
Chloroformic acid ethylester
CAEE

4. Structural and molecular formulae

4.1 Structural formula

$$\underset{Cl}{\underset{|}{}}\overset{O}{\underset{\|}{C}}-O-CH_2-CH_3$$

4.2 Molecular formula $C_3H_5ClO_2$

5. Physical and chemical properties

5.1	Molecular mass	108.52 g/mol
5.2	Melting point	$-80.6°$ C
5.3	Boiling point	$95°$ C
5.4	Density	1.135 g/cm^3 (at 20° C)
5.5	Vapour pressure	no information available
5.6	Solubility	hydrolyses in water
5.7	Solubility in organic solvents	soluble in ether, benzene, chloroform
5.8	Solubility in fat	no information available
5.9	pH-value	no information available
5.10	Conversion factor	1 ppm $\hat{=}$ 4.5 mg/m^3 air 1 mg/m^3 $\hat{=}$ 0.22 ppm (Weast 1982; Windholz, 1983)

6. Uses

Starting product for the manufacture of ore flotation agents and preservatives. Use as polymerisation catalyst (Ullmann, 1975).

7. Results of experiments

7.1 Toxicokinetics and metabolism
No information available.

7.2 Acute and subacute toxicity
The acute oral LD$_{50}$ of chloroformic acid ethyl ester was 470 mg/kg (320–690 mg/kg) for male rats and 270 mg/kg (180–400 mg/kg) for female rats. The LC$_{50}$ following a one-hour inhalation exposure was 145 ppm (140–150 ppm), or 652.5 mg/m^3 (630–675 mg/m^3) for male rats and 170 ppm (150–180 ppm), or 765 mg/m^3

(675–810 mg/m^3) for female rats. The dermal LD$_{50}$ in rabbits was 7120 mg/kg. More specific information related to these experiments are not available (Vernot et al., 1977; BASF, 1970).

In a further toxicity test, the oral LD$_{50}$ for Sprague-Dawley rats was approx. 240 mg/kg and the intraperitoneal LD$_{50}$ for NMRI mice <14.19 mg/kg (following application as a 0.5% aqueous emulsion with tragacanth) or >55.8 mg/kg (when applied as a 0.1% aqueous solution with tragacanth, 14-day follow-up period in each instance). Rats showed signs of dyspnoea, apathy, staggering and prostration on the side or stomach. On dissection, dark red discoloration of the gastric mucosa, red colouring of the intestinal contents and serous membranes were noted. The toxic symptoms in mice were dyspnoea and prostration whilst dissection revealed adhesions in the abdominal cavity (BASF, 1970).

After 3 minutes of inhaling a vapour/air mixture saturated with CAEE, 11/12 animals died. They showed an urge to escape, severe irritation of the mucous membrane and were gasping for breath. Dissection revealed an excess of blood in the lungs, associated with oedema and emphysema. Exposure of rats for 1 hour to a calculated concentration of 200 ppm caused 9 of the 10 animals to die (BASF, 1970).

In rats the repeated inhalation of 20 ppm CAEE (approx. 90 mg/m^3 air, exposure 10×6 hours) caused irritation, difficulties in breathing, a general overall poor condition and loss of body weight. Pathological examination revealed swelling and haemorrhage of the lungs. The repeated inhalation of 5 ppm (approx. 22.5 mg/m^3 air, exposure 20×6 hours) caused a retardation in body weight gain, whereas 1 ppm (approx. 4.5 mg/m^3 air, exposure 20×6 hours) had no noticeable effects (Gage, 1970).

7.3 Skin and mucous membrane effects

Undiluted CAEE applied occlusively to the skin of rabbits produced signs of corrosion after 5 minutes. After 8 days, in addition to the reddening, there were signs of oedema and necrosis (BASF, 1970).

To proof the corrosiveness of CAEE six rabbits (New Zealand, weighing 2–4 kg) were treated with the substance (0.5 ml patch) skin semi occlusive as well as occlusive for 1 and 4 hours. CAEE was corrosive independent of treatment time and procedure (Potokar et al., 1985).

The undiluted substance applied to the eyes reddening, oedema and a clouding of the cornea. During the 8-day follow-up period, purulent inflammation of the conjunctiva and corneal damage were seen (BASF, 1970).

7.4 Sensitization
No information available.

7.5 Subchronic/chronic toxicitiy
No information available

7.6 Genotoxicity

In vitro. Using the Salmonella/microsome test on strains TA 1535, TA 1537, TA 98 and TA 100, CAEE was tested for any possible mutagenic effect. The test was conducted with and without metabolic activation, using S9-mix obtained from Aroclor-induced rat livers. The test concentrations were 0, 0.05, 0.1, 0.2, 0.3 and 0.4 µl/plate with S9-mix , or 0, 0.00015, 0.001, 0.005, 0.01 and 0.015 µl/plate without S9-mix. None of the strains used within the examined concentration ranges – in the presence or absence of metabolic activation – showed any dose-related increase in the number of revertants. The maximum test concentrations were toxic to the bacteria (BASF, 1988).

In vivo. No information available.

7.7 Carcinogenicity

Groups of 50 male Sprague-Dawley rats (9 to 10 weeks old, average weight 325±16.8 g) were exposed for 30 days (6 h/day×5days/week) to 1.5 (6.75), 3.0 (13.5) or 6.0 (27) ppm (mg/m^3 air) CAEE via inhalation and than observed until they died spontaneously or sacrificed when moribund. A control group (98 animals) were exposed to air only under the same conditions. The concentration range was chosen depending of the hydrolysis rate (CAEE: $T_{1/2}$=19 minutes), because the tumor response of electrophilic substances seems to be correlated invers proportional to the hydrolysis rate. Within the section the nasal passage were flushed with 10% neutral buffered formalin; then the entire head and other organs were fixed in the same fixative. The head was then decalcified; and stepwise cross-sections were taken beginning just posterior to the nostriles and extending caudal as far as the orbit. Histological sections were taken from each lobe of the lung, trachea, larynx, liver, kidneys, testes and any other organs exhibiting gross pathology. The findings are shown in table 1.

As result in 1/50 animals (=2%, ca. 2 years old) of the highest concentration one squamous cell carcinoma on the nasal mucosa

Table 1. Exposition no. of median life animals with findings on the nasal mucosa (30 days)

	Animals	Span (days)	Rhinitis	Squamous metaplasia	Squamous cell carcinoma
Control (air)	98	613	90	9	–
CAEE					
1.5 ppm	50	576	44	13	–
3.0 ppm	50	573	43	12	–
6.0 ppm	50	617	45	15	1

was found. The rate of other tumours was comparable to the control animals as well as the animals of the lower concentrations. In the opinion of the authors the missing of a significant tumour response is related to the low concentration range, although considering the hydrolysis rate (Sellakumar et al., 1987).

CAEE was examined in three different types of studies on the mouse related to a tumorigenic efficacy (Van Duuren et al. 1987). In a 2-stage-initiation-promotion test 50 female ICR-Swiss-mice (6 to 8 weeks old) received one time 5.5 ml CAEE formulated in 0.1 ml acetone on the skin. After a period of 14 days without any application phorbol myristate acetate (PMA) as promotor was applied in the dosages of 0.0025 or 0.005 mg on the skin of the animals three times a week for the life span.

In a second long term study female mice (same age and strain) received CAEE in dosages of 3 (50 animals), 4.3 mg (30 animals) or 5.5 mg (50 animals) formulated in 0.1 ml acetone on the skin (scapular region) three times a week for the life span.

In a third long term study CAEE was injected subcutaneously once a week for the life span to female mice (same age and strain) in dosages of 0.3 or 1.1 mg formulated in 0.1 ml tricaprylin.

During the three studies body weight gain was recorded and the animals were observed daily related to tumours. Animals that became moribund or died during the treatment period were killed,

necropsied and sections were taken from the area of administration, lung, liver, kidney, spleen, colon, urinary bladder as well as organs that appeared abnormal for histopathological investigations.
The results of the three studies are shown in table 2:

Table 2. Tumour incidence in mice after administration of CAEE in differentes routes

Dosis mg/ administration	Median survival time (days)	Mice with tumours/no. mice tested	No. and tumour types at site of administration (p-value)
2-stage-test			
CAEE			
5.5	545/640	6/50	4 papilloma
			2 squamous carcinoma (p <0.05)
PMA (control)			
0.0025	455/635	0/50	0
0.005	>490/490	3/30	2 papilloma
			1 sarcoma
Repeated skin application			
CAEE			
3.0	480/660	0/50	0
4.3	>565/565	1/30	1 sarcoma (not significant)
5.5	490/660	0/50	0
Acetone (control)			
0.1 ml	545/665	0/110	0
Repeated subcutaneous application			
CAEE			
0.3	480/660	1/50	1 squamous carcinoma (not significant)
1.1	>590/590	0/30	0
Tricaprylin (control)			
0.05 ml	490/660	2/130	2 hemangioma

As result CAEE did not show a significantly higher incidence of tumours after repeated dermal or subcutaneous application for life span. The statement of the authors, that there are hints to a tumourigenic efficacy after repeated subcutaneous application is not conclusive, because only with 1/50 animals in the lower dosis (0.3 mg) a squamous carcinoma was detected whereas with 30 animals the higher dose (1.1 mg) resulted in no tumour. The authors estimated this singular tumour as not significant. On the other hand in the two stage-study there was a significant incidence of tumours, which hints at an initial efficacy of CAEE (Van Duuren et al., 1987).

7.8 Reproductive toxicity
No information available.

7.9 Effect on the immune system
No information available.

7.10 Other effects
No information available

8. Experience in humans

An employee who came into contact with CAEE during an accident at work developed an initial irritation of the eyes and of the mucosa of the respiratory tract, followed by pronounced breathing problems with cyanosis after approx. 3.5 hours. Clinical and x-ray examination indicated pulmonary oedema which was cured without complications following treatment (Bowra, 1981).

References

BASF AG
Unpublished report of 6. 2. 1970

BASF AG
Report on the study of chloroformic acid ethylester in the Ames test unpublished report of 7. 9. 1988
Commissioned by the Employment Accident Insurance Fund of the Chemical Industry

Bowra, G.T.
Delayed onset of pulmonary oedema following accidental exposure to ethyl chloroformiate
J. Soc. Occup. Med., 31, 67–68 (1981)

Gage, J.C.
The subacute inhalation toxicity of 109 industrial chemicals
Brit. J. Ind. Med., 27, 1–18 (1970)

Potokar, M., Grundler, D.J., Heusener, A., Jung, R., Mürmann, P., Schöbel, C., Suberg, H., Zechel, H.J.
Studies on the design of animal tests for the corrosiveness of industrial chemicals
Fd. Chem. Toxic., 23, 615–617 (1985)

Sellakumar, A.R., Snyder, C.A., Albert, R.E.
Inhalation carcinogenesis of various alkylating agents
J. Natl. Cancer Inst., 2, 285–289 (1987)

Ullmann (ed.)
Enzyklopädie der technischen Chemie
Verlag Chemie Weinheim, 1975, Vol. 9, p. 381

Van Duuren, B.L., Melchionne, S., Seichman, I.
Carcinogenicity of acylating agents: chronic bioassay in mice and structure activity relationships (IARC)
J. Am. Coll. Toxicol., 6, 479–487 (1987)

Vernot, E.H., MacEwen, J.D., Haun, C.C., Kinkead, E.R.
Acute toxicity and skin corrosion data for some organic and inorganic compounds and aqueous solutions
Toxicol. Appl. Pharmacol., 42, 417–423 (1977)

Weast, R.C. (ed.)
CRC Handbook of Chemistry and Physics
62nd ed. C-228
CRC-Press, Boca Raton, Florida 1981/82

Windholz, M. (ed.)
The Merck Index, 10th ed., p. 548
Merck and Co., Rahway, USA, 1983

Manganese dioxide

1. Summary and assessment

Manganese dioxide is absorbed mainly through the pulmonary system in animals and humans, but also via the gastro-intestinal tract. Accumulation of manganese following inhalation of manganese dioxide occurs chiefly in the liver, kidneys and brain, and, following oral administration, in the hair, thyroid gland, parathyroid, bones, adrenal glands und colon. Accumulation occurs in the mitochondria of each tissue, so that an impairment of enzyme functioning (which is important for oxidation) must be assumed. An inhibition of specific enzymes (succinic acid dehydrogenase, cytochrome oxidase, rhodanase and lactic acid dehydrogenase) has been demonstrated in animals following the intake of manganese dioxide. It has been suggested that an inhibition of the respiratory enzyme system can occur in areas of the brain sensitive to oxygen deficiency, following administration of manganese dioxide. Excretion of the administered manganese dioxide occurs essentially via the bile in the faeces (measured as manganese).

There are no relevant data available for assessing the acute oral toxicity of manganese dioxide in animals. Studies of acute toxicity following administration subcutaneously (in the mouse) and intravenously (in the rabbit) indicate a moderate degree of toxicity. However, repeated administration of sub-lethal doses (orally 500 mg/kg; on inhalation up to 219 mg/m^3 of air) suggests that the toxicity is only slight on single intake. So far, experience of the effects on humans indicates that acute intoxication with manganese dioxide does not occur to any great extent.

Rats exposed for 10 days to concentrations of 68, 130 and 219 mg manganese dioxide dust/m^3 of air, have shown dose-dependent focal pneumonitis and interstitial hypercellularity at the lowest concentration and, at the higher concentrations, a diffuse pneumonitis and matured granulomas with, among other things, histiocytes, fibro-connective tissue and foreign-body multinucleate giant cells. Experience of exposure in humans also points to this pattern of damage. Inflammation of the upper respiratory tract and severe cases of pneumonia have been observed in persons with prolonged exposure to manganese dioxide dust or vapour. This manganese pneumonia is similar to lobular pneumonia, but is minimally responsive to antibiotic treatment and has a high mortality rate.

Inhalation studies on mice exposed to manganese dioxide dust for up to 32 weeks (concentrations of 49.1 and 88.9 mg/m^3 of air) have shown changes in behaviour and learning capacity, but these effects are no longer observable during the subsequent manganese dioxide-free period (6 weeks) and thus appear to be completely reversible. Mice which have been administered corresponding manganese dioxide concentrations in the food over a similar period of time show no evidence of damage.

In a reproductive study of female mice exposed before conception to manganese dioxide dust, it is found that the resulting litter has a lower growth rate and shows abnormal behavior (in the form of reduced activity).

In humans exposed over prolonged periods to manganese dioxide dust or vapour, a morbid condition develops, the symptoms of which are largely identical to those of Parkinsonism. The focal point of damage is the brain stem with expecially pronounced degeneration of the striatum and pallidum. In addition to central nervous system damage, peripheral nerves can also be affected. Other types of syndrome, resembling bulbar paralysis, amyotrophic lateral sclerosis or multiple sclerosis have also been described. Indulgence in alcohol appears to play an important role in the degree of severity of the resulting illness.

No increase in the frequency of leukaemia, lymphoma or fibrocarcinoma (at the site of injection) have been observed following repeated intramuscular injection of manganese dioxide in rats and mice; insofar there is no evidence of a carcinogenic action.

A carcinogenic effect of manganese dioxide on humans has not been reported.

In summary, it appears that prolonged exposure to manganese dioxide in particular via the respiratory system, leads to definite, partially irreversible damage both to the respiratory system and to the central nervous system in animals and humans. An effect on the offspring cannot be excluded.

2. Name of substance

2.1	Usual name	Manganese dioxide
2.2	IUPAC name	Manganese dioxide
2.3	Cas-No.	1313-13-9

Manganese dioxide

3. Synonyms

Common and trade names — Braunstein (Pyrolusite = manganese mineral with 70 to 80% manganese dioxide)

4. Structural and molecular formulae

4.1 Structural formula — MnO_2, rutile structure and other polymorphics

4.2 Molecular formula — MnO_2

5. Physical and chemical properties

5.1 Molecular mass — 86.94 g/mol
5.2 Melting point — 535° C (Neumüller, 1985)
5.3 Boiling point — –
5.4 Density — 5.026 g/cm^3 (at 20° C) (Neumüller, 1985)
5.5 Vapour pressure — no information available
5.6 Solubility in water — insoluble (Neumüller, 1985)
5.7 Solubility in organic solvents — –
5.8 Solubility in fat — –
5.9 pH-Value — –
5.10 Conversion factors — 1 ppm $\hat{=}$ 3.607 mg/m^3
1 mg/m^3 $\hat{=}$ 0.277 ppm

6. Uses

In the manufacture of manganese, ferromanganese and other manganese compounds; in electrical engineering for batteries; in the glass, porcelain and ceramics industry for dyeing and bleaching; as a siccative for oils and varnishes; filler dye for rubber; for metal burnishing; for the de-ironing of water; as a catalyst (Neumüller, 1985; Windholz, 1983).

7. Results of experiments

7.1 Toxicokinetics and metabolism

Manganese is absorbed both via the lungs and via the gastrointestinal tract. Following repeated inhalation exposure, a raised manganese content (no data on values) was obtained in the blood, liver, kidneys, lungs, cerebellum, brain stem and testes of mice, the level decreasing again (liver excepted) during further exposure (Morganti et al., 1985).

Repeated oral dosing (500 mg manganese dioxide/kg body weight by stomach tube, weekly for 6 months) led to raised manganese content (no further details) in the hair, thyroid, parathyroids, bones, adrenals and the large intestine of male rats. The rate of accumulation in the liver was increased 1000-fold by comparison with untreated control animals (Sakamoto, 1982).

Manganese dioxide became intermediately bound to protein. Excretion took place essentially in the faeces via the bile and intestine (Taeger, 1941; Southern, 1983). Manganese dioxide accumulated chiefly in the mitochondria, leading to the postulation of an inhibition of the enzymes fundamentally important in oxidation processes (Jonderko and Zahorski, 1972).

7.2 Acute and subacute toxicity

No results are available relating to investigations on acute toxicity following oral, dermal or inhalation exposure in animals. A single subcutaneous dose given to mice produced an LD_{50} of 422 mg manganese dioxide/kg body weight. Following a single intravenous injection in rabbits, an LDLo of 45 mg manganese dioxide/kg body weight was established (Merck Index, 1983).

In a subacute inhalation study on rats of both sexes (3 animals/concentration) that were exposed for 6 hours/day for 10 days to 68, 130 and 219 mg manganese dioxide/m^3 air, in the form of a dust aerosol, there were indications of pneumonitis and, at the higher concentrations, granulomatous changes (Shiotsuka, 1984).

7.3 Skin and mucous membrane effects
No information available.

7.4 Sensitization
No information available.

7.5 Subchronic-chronic toxicity
No information available.

Manganese dioxide 331

7.6 Genotoxicity
In vitro. No information available.
In vivo. No information available.

7.7 Carcinogenicity
25 F344 rats of both sexes and 25 female Swiss-Albino mice received intramuscular injections of manganese dioxide fine powder, formulated in 0.2 ml Trioctanoin. The rats received 10 mg manganese dioxide/animal once a month for 9 months, while the mice received 3 or 5 mg/animal once a month for 6 months. The animals were weighed each week for the first month and then each month. Once a week all the animals were palpated. All the animals were dissected at the end of the experiment, and the lungs, livers and any portions of the tissues with conspicuous changes were removed and fixed in formalin (10%). The dissected material was stained with haematoxylin-eosin and examined histologically. The results obtained are shown in table 1 (see page 324). As can be seen from the table, in comparison with controls there was no increase in the incidences of leukaemias, lymphomas or fibrosarcomas (rats only) in rats or mice following intramuscular injection of manganese dioxide (Furst, 1978).

Sundermann (1983) was similarly able to show in male Fischer rats that following intramuscular injection of manganese dioxide (at a dose calculated as being equivalent to 1 mg Mn), no fibrosarcomas had occurred at the site of injection after 100 weeks (no further details).

7.8 Reproductive toxicity
Female mice were exposed to manganese dioxide dust (12 weeks at approx. 48 mg/m^3 air, then at approx. 85 mg/m^3 air) or to filtered air (7 hours a day, 5 days a week, for 16 weeks prior to conception). On the 1st day of gestation half of each group was exposed either to manganese dioxide dust or to filtered air only, continuing up to the 17th day of gestation. In order to exclude effects of a prenatal exposure to manganese dioxide prior to postnatal uptake of manganese dioxide by suckling mice via the milk, untreated mothers (wet nurses) were used in the corresponding groups. The behaviour of the neonates was tested in the activity field or circular-track test. By comparison with the control mothers, the number of young per litter of the mothers exposed prior to conception was increased significantly (p <0.05). Prenatal exposure led, however, to reduced activity behaviour of the neonates and to retarded growth

Table 1. Carcinogenic effects in rats and mice after chronic intramuscular injection of MnO_2

Rats	Weight at the end of the experiment in g (mean value)		Total dose/ animal	Lymphomas and leukaemias		Fibrosarcomas at the injection site	
	Male	Female		Male	Female	Male	Female
Trioctanoin-controls	382	269	2.4 ml	0/25	4/25	0/25	0/25
MnO_2-animals	358	260	90 mg	0/25	3/25	0/25	0/25

Mice (female)	Weight at the end of the experiment in g (mean value)	Total dose/ animal	Leukaemias	Lymphomas
0-controls	33	0	5/25	1/25
Trioctanoin-controls	35	2.4 ml	2/25	0/25
	33	18 mg	4/25	1/25
MnO_2-animals	36	30 mg	1/25	2/25

(7% less than in the control) that was observable up to 45 days post partum. Neonates that were suckled by mothers exposed to filtered air prior to conception showed an increased activity score on the 12th day by comparison with neonates suckled by mothers exposed to filtered air prior to conception and thereafter to manganese dioxide. Frequency of movement, exploratory behaviour and scores from activity behaviour (tested in the open field test, rotating-bar test, and circular-track test) were reduced in the sexually mature young animals that had taken up manganese dioxide either in utero or via the milk. Independently of whether or not there was exposure to manganese dioxide in utero, sexually mature young animals that were suckled by mothers exposed to manganese dioxide had a significantly lower manganese content in the mitochondria of the cerebellum and the brain stem (reduced by approx. 50%, p <0.01; Lown et al., 1984).

7.9 Effects on the immune system
No information available.

7.10 Other effects

Manganese dioxide exhibited an inhibitory effect on specific enzyme activities (succinic acid dehydrogenase, cytochrome oxidase, rhodanase, lactic acid dehydrogenase; Jonderko, 1972).

96 adult male mice (ICR, Swiss-strain, approx. 3 months old) underwent whole-body exposure to manganese dioxide dusts or to filtered air (control) for 32 weeks. In the first 12 weeks the manganese dioxide concentration was approx. 49 mg manganese dioxide/m^3 air, while from the 13th week it was approx. 85 mg manganese dioxide/m^3 air. Treated and control animals were subsequently exposed to the control conditions for a further 6 weeks. Measurement of manganese in the tissues began in the 17th week together with behaviour learning tests, and 8 animals were sampled at random from the groups each month in order to carry out these investigations. The remaining animals continued to be subject to the test program. The result with regard to distribution in the tissues was described in 7.1; no pathological changes were found. Exposure to manganese dioxide nevertheless led to significant differences in the behaviour and learning ability of the animals (increased activity, longer latency of passive avoidance tested in the open field test; circular-track test, rotating-bar test, and two-chamber test system). These changes subsided very rapidly in the manganese dioxide-free exposure period (6 weeks) and were thus reversible (Morganti, 1985).

Animals exposed to similar manganese dioxide concentrations in the diet (1 mg/g food) over a comparable experimental period (30 weeks) showed no significantly altered modes of behaviour (Morganti et al., 1985).

8. Experience in humans

Cases of acute poisoning by manganese dioxide are virtually non-existent (Moeschlin, 1980).
Poisoning is possible as a result of the inhalation of manganese vapours or dusts. Chronic poisoning is chiefly evoked by fairly long-term uptake of manganese dioxide dusts or vapours. Accumulation occurs essentially in the liver and kidneys, and also in the central nervous system (plexus chorioideus). Excretion takes place mainly via the liver and bile in the faeces (Bittersohl et al., 1972; Taeger, 1941).
Two clinical pictures stand out:
1. Manganese pneumonia
2. Manganism
Manganese pneumonia resulting from fairly long-term dust inhalation follows indirectly from the inflammation of the upper respiratory tract, a primary effect of which is markedly to reduce the defences of the organism against the ubiquitous causal organisms of pneumonia (Sturm, 1972; Taeger, 1941).
Roentgenologically, and also anatomically and pathologically, manganese pneumonia is comparable with a lobular pneumonia, but its course is considerably more serious. It is almost completely resistant to antibiotics and shows a high mortality rate (Bittersohl et al., 1972).
The clinical picture of manganism corresponds largely with Parkinsonism in its symptoms. The focal point of damage is the brain stem, the striatum and pallidum being particularly affected. Apart from the central nervous system damage, peripheral nerves may also be involved (signs: pro- and retropulsion). Other types of clinical course resemble bulbar paralysis, amyotrophic lateral sclerosis, and multiple sclerosis. A fairly long period of exposure (uptake over 2 to several years) is generally required prior to the appearance of symptoms. Cases of poisoning have, however, occurred after exposure for no more than a few months (Bittersohl et al., 1972; Henschler, 1980; Moeschlin, 1980).

Results of an epidemiological study on 141 workers in a manganese salt and oxide plant indicated preclinical disturbances (chiefly impairment of lung function and of the central nervous system; Roels, 1985).

References

Bittersohl, G., Ehrhardt, W., Grund, W., Grunewald, A.
Chemische Schädigungen
in: F. Koelsch „Handbuch der Berufserkrankungen", p. 173–176
VEB Gustav Fischer Verlag, Jena (1972)

Furst, A.
Tumorigenic effect of an organomanganese compound on F344 rats and Swiss Albino mice: brief communication
J. Natl. Cancer Inst., 60(5), 1171–1173 (1978)

Henschler, D.
Wichtige Gifte und Vergiftungen
in: W. Forth, D. Henschler, W. Rummel
Allgemeine und spezielle Pharmakologie und Toxikologie, p. 603
B.I. Wissenschaftsverlag, Mannheim (1980)

Jonderko
cited from Bittersohl et al.
in: F. Koelsch „Handbuch der Berufserkrankungen", p. 173–176
VEB Gustav Fischer Verlag, Jena (1972)

Lown, B.A., Morganti, J.B., D'Agostino, R., Stineman, C.H., Massaro, E.J.
Effects on the postnatal development of the mouse of preconception, postconception and/or suckling exposure to manganese via maternal inhalation exposure to MnO2 dust
Neurotoxicology, 5, 119–131 (1984)

Moeschlin, S.
Vergiftungen durch anorganische Stoffe
in: S. Moeschlin
Klinik und Therapie der Vergiftungen, p. 192–194
Georg Thieme Verlag, Stuttgart, New York (1980)

Morganti, J.B.
Uptake, distribution and behavioral effects of inhalation exposure to Manganese (MnO2) in the adult mouse
Neurotoxicology, 6, 1–16 (1985)

Neumüller, O.A.
Römpp Chemielexikon, 8th Edition, p. 2482
Franckh'sche Verlagshandlung, Stuttgart (1985)

Pazynich, V.M.
Functional state of the central nervous system in animals in the presence of manganese dioxide aerosols
Gigiena Naselen Mest. Kiev, JSS, 24, 87–90 (1985)

Roels, H.
Preclinical toxic effects of manganese in workers from a manganese salts and oxides producing plant
Sci. Total Environ., 42(1–2), 201–206 (1985)

Sakamoto, M.
Effect of insoluble manganese compounds on the whole body
1. Rats given the compounds through a stomach tube
Hokuriku Hoshu Eisei Gakkaishi, 9(1), 22–26 (1982)

Shiotsuka, R.N.
Inhalation toxicity of MnO2 and a magnesium oxide-manganese dioxide mixture
U.S. Army Medical Research and Development Command
Govt. Accession No. AD-148868 (1984)

Snella, M.C.
Manganese dioxide induces alveolar macrophage chemotaxis for neutrophils in vitro
Toxicology, 34(2), 153–160 (1985)

Southern, L.L.
Excess manganese ingestion in the chick
Poult. Sci., 62(4), 642–646 (1983)

Sturm, W.
Luftwege und Lungen
in: F. Koelsch „Handbuch der Berufserkrankungen", p. 682
VEB Gustav Fischer Verlag, Jena (1972)

Sundermann, F.W. Jr.
Effects of manganese compounds on carcinogenicity of Nickel subsulfide in rats
Carcinogenesis 4(A), 461–465 (1983)

Taeger, H.
Die Klinik der entschädigungspflichtigen Berufskrankheiten
p. 116, Springer, Berlin (1941)
cited from S. Moeschlin in: Klinik und Therapie der Vergiftungen, p. 192–194
Georg Thieme Verlag, Stuttgart, New York (1986)

Windholz, M. (ed.)
The Merck Index, 10th ed., p. 817
Merck and Co., Rahway, USA (1983)

Zahorski
cited from Bittersohl et al.
in: F. Koelsch „Handbuch der Berufserkrankungen", p. 173–176
VEB Gustav Fischer Verlag, Jena (1972)

Chemical index

Adipic acid dinitrile 252
Adiponitrile 252
Adipyl dinitrile 252
AH 182
2-Amino-1-methoxy-4-nitrobenzene 155
1-Amino-4-chloro-2-nitrobenzene 75
1-Amino-4-methoxy-2-methylbenzene 173
2-Amino-5-chloro-nitrobenzene 75
Ammonyx 200 312
Azoamine Scarlet K 155
Azoic Diazo Component 13, Base 155

1,4-Benzene dicarbonic acid dimethyl ester 266
1,4-Benzene dicarboxylic acid dimethyl ester 267
N,N'-Bis(1-methylpropyl)-1,4-benzenediamine 166
N,N'-Bis(1-methylpropyl)-1,4-diaminobenzene 166
BLO 134
Braunstein 329
Brecolane NDG 219
Butane dinitrile 252
1,4-Butanolide 134
Butyl phosphate 298
2-Butyne-1,4-diol 207
Butynediol 207
Butyric acid 4-hydroxy-, gamma-lactone 134
Butyric acid lactone 134

4-Butyrolactone 134
Butyryl lactone 134

CAEE 319
CAS-No.
78-83-1 44
79-07-2 59
89-63-4 75
96-45-7 83
96-48-0 134
99-59-2 155
101-96-2 166
102-50-1 173
104-76-7 182
110-65-6 207
111-46-6 219
111-69-3 252
120-61-6 266
123-51-3 284
126-73-8 298
138-24-9 311
541-41-3 319
1313-13-9 328
Celluphos 4 298
4-Chloro-2-nitroaniline 75
p-Chloro-o-nitroaniline 75
Chloroacetamide 59
α-Chloroacetamide 60
2-Chloroacetamide 59
Chloroacetic acid amide 60
Chloroformic acid ethyl ester 319
CNA 75
m-Cresidine 173
Deactivator E 219
Deactivator H 219

Chemical index

DEG 219
N,N'-Di-sec.-butyl-p-phenylenediamine 166
DICOL 219
1,4-Dicyanobutane 252
Diethylene glycol 219
Diglycol 219
Dihydro-2(3H)-furanone 133
4,5-Dihydro-2-mercaptoimidazole 83
beta,beta'-Dihydroxydiethyl ether 219
Dimethyl terephthalate 266
Dimethyl-1,4-benzene dicarboxylate 267
Dimethyl-p-phthalate 267
But-2-yne-1,4-diol 207
1,4-Dioxy-butyne-2 207
Disflamol 1 TB 298
Dissolvand APU 219
DMT 267
DuPont Gasoline Antioxidant No. 22 166

Ethanol, 2,2'-oxydi- 219
2-Ethyl-n-hexyl-alcohol 182
Ethylchlorocarbonate 319
Ethylchloroformiate 319
Ethylchloromethanate 319
Ethylene diglycol 219
Ethylene thiourea 83
Ethylenethiourea 83
N,N'-Ethylenethiourea 83
1,3-Ethylenethiourea 83
2-Ethylhexan-1-ol 182
2-Ethylhexanol-1 182
2-Ethylhexanol 182
2-Ethylhexyl alcohol 182
ETU 83
gamma-BL 134

gamma-Butyrolactone 133
GBL 134
Glycol ether 219
Glycol ethyl ether 219
Golpanol 207

Hexanedioic acid dinitrile 252
4-Hydroxybutanoic acid lactone 134
4-Hydroxybutanoic acid, gamma-lactone 134
gamma-Hydroxybutyric acid cyclic ester 134
gamma-Hydroxybutyric acid lactone 134
4-Hydroxybutyric acid lactone 134
4-Hydroxybutyric acid, gamma-lactone 134
Bis(2-hydroxyethyl)ether 219
1-Hydroxymethylpropane 44

Imidazolidinethione 83
2-Imidazolidinethione 83
Imidazoline-2(3H)-thione 83
Imidazoline-2-thiol 83
Isoamyl alcohol 284
Isobutanol 44
Isobutyl alcohol 44
Isobutylcarbinol 284
Isoctanol 182
Isopentyl alcohol 284

Kerobit BPD 166
Korantin BH 207, 209

Manganese dioxide 328
2-Mercapto-2-imidazoline 83
Mercaptoimidazoline 83
Mercazin 1 NA 22 83
4-Methoxy-2-methylaniline 173

Chemical index 341

4-Methoxy-2-methylbenzenamine 174
2-Methoxy-5-nitroaniline 155
2-Methoxy-5-nitrobenzenamine 155
Methyl-4-carbomethoxybenzoate 267
2-Methyl-4-methoxyaniline 174
2-Methyl-p-anisidine 173
Bis-oxy methylacetylene 207
3-Methylbutanol-1 284
2-Methylpropan-1-ol 44
2-Methylpropane-1-ol 44
2-Methylpropanol-1 44
2-Methylpropyl alcohol 44
Microcide Mergal AF 60

NCI-C55878 134
5-Nitroanisidine 155
5-Nitro-ortho-anisidine 155

Octanol, technical 182
3-Oxa-1,5-pentanediol 219
3-Oxapentane-1,5-diol 219
2,2'-Oxybisethanol 220
2,2'-Oxydiethanol TL4N 220

Pennac CRA 83
Phenyltrimethylammonium chloride 311

Phosphoric acid tri-n-butylester 298
Pyrolusite 329

Rhenogram ETU 83
Rhodanin S 62 83

Soxinol 22 83

TBP 298
Tenamene 2 166
Terephthalic acid dimethyl ester 267
Tetramethylene cyanide 252
Tetramethylene dicyanide 252
Thiourea, N,N'-(1,2-ethandiyl) 83
Tributyl phosphate 298
Trimethylanilinium chloride 312
N,N,N-Trimethylbenzen-ammonium chloride 311
Trimethylphenyl ammonium chloride 311
Tri-n-butyl phosphate 298

Vulcacit NPV/C 83

Warecure C 83